Advances in
ARTIFICIAL INTELLIGENCE

Advances in
ARTIFICIAL INTELLIGENCE

Edited by

John Hallam,
Department of Artificial Intelligence,
University of Edinburgh

and

Chris Mellish,
Cognitive Studies Programme,
University of Sussex

Proceedings of the 1987 AISB Conference,
University of Edinburgh, 6 – 10 April 1987.

JOHN WILEY & SONS
Chichester · New York · Brisbane · Toronto · Singapore

British Library Cataloguing in Publication Data available

Printed and bound in Great Britain

Acknowledgements

This small volume contains the papers presented at AISB-87, the AISB conference held at the University of Edinburgh in April 1987. The Society for the Study of Artificial Intelligence and the Simulation of Behaviour (SS AISB) was founded in 1964 and acts as a focal point for news of Artificial Intelligence research and experiments in the simulation of behaviour. Membership is about 1000, drawn from both Universities and Industry.

AISB-87 Programme Committee:

Tony Cohn	Mark Elsom-Cook	Bob Fisher
Alison Kidd	Rudi Lutz	Chris Mellish

AISB-87 Tutorial Organiser:

Masoud Yazdani

AISB-87 Local Arrangements Organisers:

John Hallam   Fiona Gordon

AISB-87 Referees:

Andrew Blake	Mike Brayshaw	Derek Brough
Alan Bundy	Bernard Buxton	Mike Clarke
Ian Craig	Jon Cunningham	Mark Drummond
Ben Du Boulay	Marc Eisenstadt	Roger Evans
Rick Evertsz	Tony Hasemer	David Hogg
Simon Holland	Steve Isard	George Kiss
Ewan Klein	Simon Love	John Lumley
Chris Malcolm	Abe Mamdani	Chris Miller
Alex Morrison	Simon Nuttall	Steve Pulman
Allan Ramsay	Steve Reeves	Graeme Ritchie
Peter Ross	Nigel Shadbolt	Bill Sharpe
Mike Sharples	Karen Sparck-Jones	Phil Stenton
Spud Tait	Josie Taylor	Chris Thornton
Stephen Todd	Mike Uschold	Andrew Wallace
Lincoln Wallen	Bob Welham	Andrew Whiter
Stuart Wooller	David Young	Richard Young

Foreword

Although the "AI Business" has grown and will continue to grow at great pace, the fact remains that most current practical applications of Artificial Intelligence are based on technical advances that were made more than ten years ago. The financial rewards (and the government grants available) for the applied researcher are considerable, and so, although the number of people actively working in AI is undoubtedly growing rapidly, there is a natural tendency for people to move away from theoretical, long-term research. As a result of this, the question naturally arises: will AI be able to provide the insights necessary to support a next generation of "expert systems" for application in ten years' (or so) time? In terms of applications, this amounts to the question as to whether the future growth of the "AI Business" will be through selling "more of the same" to more consumers on cheaper and more powerful hardware, or whether it will rest on new and improved kinds of products that will revolutionise our views on the roles of people and computers in Society. In terms of scientific progress, it amounts to the question as to whether AI, having found some initial applications, will stagnate as an intellectual discipline, or whether genuine theoretical advances will continue to be made.

The tone and quality of the papers included in this book indicate that healthy fundamental research is continuing in AI and give cause for optimism about the future. The problems that researchers are addressing are the classical problems of AI—how can we produce intelligent behaviour by constructing systems that represent and reason about the world? The particular areas that have attracted most attention here are representing and reasoning about time, physical objects, belief and linguistic structure. Understanding these areas has occupied man from the very beginning, and we are certainly nowhere near having a complete account for any of them in current AI work (or indeed in other disciplines). The way in which the papers presented here build and improve on previous work leads one to believe that real progress is, however, being made. There seems to be a new seriousness and rigour in current AI work, often associated with the increased use of formal techniques. As a result, it is generally easier in the 1980s to assess the capabilities and limitations of a piece of work than it was in the 1970s.

Of course, for representation systems to be of any use in AI, they must support computationally feasible reasoning processes of some kind. The standard problem for a reasoning system is search, and we have several papers looking at the problem of search control in various kinds of reasoning systems. In addition, we can always gain insight from the careful observation of how human beings solve problems, both when they are successful and when they are not. It is thus only right that we have some papers on human problem solving in this collection.

It is interesting that so little of the work described here is influenced by the "parallel distributed processing" or "connectionist" schools of thought so active in the USA. Researchers in these traditions would probably argue that much of the work on explicit, local representations (and the associated reasoning mechanisms) presented here is fundamentally misguided. Why do we hear so little of "connectionism" in our collection of papers? It is undoubtedly partly because the papers here are drawn mainly from Europe. European AI has always emphasised traditional formal approaches, such as theorem proving, in contrast to approaches that might amount to unprincipled proceduralism. In addition, European researchers rarely have the prodigious computer resources needed to do serious "connectionist" experiments. Finally, although "connectionism" provides some inspiring metaphors and insights, it still has to prove itself as a methodology for building practical intelligent systems or for producing interesting scientific theories.

Whereas many Artificial Intelligence conferences accept a large number of papers and make extensive use of parallel sessions, AISB-87 has been designed to be a small conference with no parallel sessions. The organisers hope in this way to make it easier for researchers to make links between different areas of AI research, and indeed for some of the links with other Cognitive Science researchers to become more apparent. The sections of this book, which reflect the sessions of the conference therefore do not have familiar titles like "computer vision" and "theorem proving". We have tried as far as possible to exploit common themes in the accepted papers, rather than attempt to force them into traditional categories.

We would like to thank the contributors to this volume and also all the many people who have helped to make the AISB-87 conference come to exist.

Chris Mellish, *University of Sussex*
John Hallam, *University of Edinburgh*

Contents

Contributors

Afzal Ballim, *Rio Grande Research Corridor, Computing Research Laboratory, New Mexico State University, Box 3CRL Las Cruces, NM 88003, USA*

Paul Brna, Alan Bundy, Helen Pain and Liam Lynch, *Department of Artificial Intelligence, University of Edinburgh, 80 South Bridge, Edinburgh EH1 1HN*

Andy Clark, *Cognitive Studies Programme, Arts Building, University of Sussex, Falmer, Brighton BN1 9QN*

Martin Conway and Hank Kahney, *MRC Applied Psychology Unit, 15 Chaucer Road, Cambridge CB2 2EF*

Roddy Cowie, *Department of Psychology, The Queen's University of Belfast, Belfast BT7 1NN, Northern Ireland*

Roger Evans, *Cognitive Studies Programme, Arts Building, University of Sussex, Falmer, Brighton BN1 9QN*

David Frost, *Biomedical Computing Unit, Imperial Cancer Research Fund Laboratories, PO Box No 123, Lincoln's Inn Fields, London WC2A 3PX*

Patrick Hayes, *Schlumberger Palo Alto Research, 3340 Hillview Avenue, Palo Alto, California 94304, USA*

Roman Jansen-Winkeln, *c/o Prof Dr W Wahlster, Universitaet des Saarlandes, FB 10—Informatik, Im Stadtwald 15, D – 6600 Saarbrucken 11, Fed Rep of Germany*

John Kelly, *Department of Computer Science, University College Dublin, Belfield, Dublin 4, Republic of Ireland*

Mauro di Manzo and Emanuele Trucco, *D.I.S.T.—University of Genoa, via all Opera Pia 11a, 16145 Genova, Italy*

Tony Morgan, *University of Cambridge Computer Laboratory, Corn Exchange Street, Cambridge CB2 3QG*

Alexander Nakhimovsky, *Department of Computer Science, Colgate University, Hamilton, New York 13346, USA*

Alan Ramsay, *Cognitive Studies Programme, Arts Building, University of Sussex, Falmer, Brighton BN1 9QN*

Steve Reeves, *Dept of Computer Science & Statistics, Queen Mary College, Mile End Road, London E1 4NS*

Han Reichgelt, *Department of Artificial Intelligence, University of Edinburgh, 80 South Bridge, Edinburgh EH1 1HN, Scotland*

David Rumelhart, *Programme in Cognitive Science, University of California San Diego, La Jolla, California 92093, USA*

Derek Sleeman, *Department of Computer Science, University of Aberdeen, Kings College, Old Aberdeen AB9 2UB*

Sam Steel, *Department of Computer Science, University of Essex, Wivenhoe Park, Colchester CO4 3SQ*

Sam Steel and Anne de Roeck, *Department of Computer Science, University of Essex, Wivenhoe Park, Colchester CO4 3SQ*

Edward Tsang, *Department of Computer Science, University of Essex, Wivenhoe Park, Colchester CO4 3SQ*

Lincoln Wallen and Gregory Wilson, *Department of Artificial Intelligence, University of Edinburgh, 80 South Bridge, Edinburgh EH1 1HN*

Bo Zhang and Ling Zhang, *Department of Computer Science, Tsinghua University, Beijing, China*

Philosophical Issues

CONNECTIONISM AND COGNITIVE SCIENCE   *

Andy Clark

Cognitive Studies Programme, University of Sussex,
Brighton, U.K.

ABSTRACT

How does work in the connectionist tradition relate to work in
conventional AI? And what, if anything, can connectionism tell us
about the nature of human cognition? There are, as yet, no clear
answers to either question. The present paper scouts the conceptual
terrain and isolates some of the special features and difficulties
of the connectionist approach. A standard polarisation ('either
connectionism is a mere implementation theory or it reveals the deep
structure of human cognition') is resisted and an alternative
briefly developed.

1. INTRODUCTION; A CONSTELLATION OF NEW IDEAS?

Connectionism (aka parallel distributed processing) is a rising star
in the firmament of cognitive science. It presents, some say, a
'whole new constellation of assumptions about the nature of cognitive
processes' (McClelland and Rumelhart,1985 p.184). According to some
of its adherents, more conventional accounts involving serial
processes acting on atomistic knowledge, may be doomed to the status
of 'convenient, approximate descriptions of the underlying structure
of memory and thought' (McClelland and Rumelhart, 1985a p.196).
Connectionism, on this radical view, unearths the fine-grained truth
about cognitive processes. There are those who are less enthusiastic.
For them, connectionism has roughly the status of an implementation
theory. It adds nothing to our understanding of the psychological
or computational facts (see Broadbent, 1985; Pylyshyn, 1984 p.215).
Instead, we learn something about how the nervous system instantiates
such facts.

In the present paper I consider the evidence for each position (the
radical and the unenthusiastic) and find it wanting. It is, for

  *   This paper owes much to a seminar and discussion with G.Hinton
      at the University of Sussex, April 1986. I have also benefitted
      from reading S.W. Draper's draft paper 'Does connectionism
      constitute a paradigm revolution?', and from a seminar on this
      topic attended by S.W.Draper, J.Stone, S.Torrance, C.O'Malley,
      B.Whitby and others. Special thanks to J.Stone, S.W.Draper
      and A.Sloman for their detailed comments on an earlier draft.

3

example, an unargued assumption of each side that there will exist a <u>uniform</u> relation between conventional and connectionist models. But it may well be that connectionism reveals the fine-grained psychological and computational structure of <u>some</u> aspects of mind while being a mere implementation theory for <u>other</u> aspects (which may depend on a (possibly virtual) von Neumann architecture).

The structure of the paper is as follows. I begin (section II) by sketching some common features of connectionist models of cognitive processes. I go on (section III) to focus on two particularly radical features of such models viz. their use of superposed representations and their powerful learning capacities. The next section (IV) considers the extent to which such models may be seen as psychologically informative and the common criticism of connectionism as a 'mere implementation theory'. I end (V) by making some exploratory comments on possible relations between connectionist models and conventional AI.

II. CONNECTIONISM – A SKETCH.

The musician's talent, it is often said, lies not in playing the notes but in <u>spacing</u> them. It is the <u>silences</u> which make the great musician great. As it is with music, so with connectionism. The power of a connectionist architecture lies not in the individual units but in the subtly crafted connections between them. A connectionist model which has aroused much interest is the Boltzmann machine. Before describing this I should point out that the algorithmic form of connectionist models varies quite extensively. This paper focusses on sets of emergent behaviours common to a large proportion of connectionist models of cognitive (or allegedly cognitive – see section 5.1) processes. The examples I use are therefore designed to bring out common, but by no means prototypical, features of this general class of machines.

The Boltzmann machine, (as depicted in Ackley, Hinton and Sejnowski, 1985), is composed of many simple units operating in parallel and connected to their neighbours by bi-directional links. The links are weighted either positively or negatively. Suppose we conceive each individual unit as representing a primitive hypothesis about some target domain. The unit fires when it 'believes' the hypothesis is true. Two units which stand for contradictory hypotheses may then be linked by a negatively weighted connection. If one fires, it will tend to inhibit the other. Mutually supporting hypotheses may be linked by a positively weighted (excitory) connection. The links thus allow the individual units to excite and inhibit each other in a systematic manner. The state of a unit at a given time will depend, in part, on the state of all the units to which it is linked. And those units, in turn, will be influenced by all the units to which they are linked. An iterative process of mutual adjustment of response ensues until a 'communal decision' is reached. This process is sometimes termed <u>relaxation</u>.

To take an example, suppose the initial state of activity of the

units is a direct function of a raw perceptual input in the form of
an intensity array. Each unit (this is a simplification – groups of
units would almost certainly be required) is primed to respond to one
kind of feature in such an array. It must also, however, listen to
the 'opinions' of any units to which it is connected. In due course
the global network should relax into a communal, internally
consistent decision. This will amount to an interpretation of the
intensity array in terms ultimately of a 3D scene. If the connections
between the units have been well chosen, the system should (often)
get it right. See e.g. Ballard, Hinton and Sejnowski,1983 p.21.

The essential point to note, then, is that connectionist machines
(as I shall use the term) are not just vast parallel processors.
Parallelism alone is not enough. Rather, what counts is a process
of co-operative group decision. Co-operative algorithms work to
achieve (by a process of iterative adjustment) an interpretation
which respects constraints between neighbouring elements.
Co-operation is therefore local, whereas the emergent order (the
simultaneous satisfaction of a large number of such constraints) is
global. A homely example (which I first heard from J.Stone) is that
of the open market place. Here global patterns of supply and demand
are established by local interactions of buying and selling. Overall
knowledge of demand is thus distributed amongst buyers and overall
control of supply amongst suppliers.

A second example of a connectionist model is McClelland and
Rumelhart's interactive-activation model of word recognition. This
employs a connectionist architecture to embody relations of mutual
support and constraint between letters in the process of word
recognition. Thus units responding to features of a given letter in
the first position in a word will excite units for that whole letter
and they in turn will excite units for words beginning with that
letter. Mutually exclusive units (e.g. those which pick out features
suggesting different letters, or which pick out different initial
letters) will be set up so as to inhibit one another. Connectionist
models, however, are by no means _essential_ for the achievement of
such tasks. In principle, serial algorithms and explicit constraints
on interpretation could achieve these goals (and do so _without_
simulating the connectionist architecture itself). What, then, is
the special interest of the connectionist's proposals?

Two immediate attractions are the _speed_ of parallel co-operative
algorithms relative to certain kinds of task, and their _robustness_
in coping with local hardware damage. The first of these, the time
constraint, is crucial. Take human vision for example. This seems
to require quite massive amounts of computation. Yet given what we
know of the speed and architecture of the human brain, we are far
too slow and ponderous to produce results as fast as we do _unless_
we abandon the assumption of serial processing. Deploy parallel
computing techniques (in which many simple units work
simultaneously on different aspects of the problem and communicate
amongst themselves) and the problem becomes tractable. The speed
of human visual response, then, provides one very strong reason to
believe that parallel algorithms are involved (see e.g. Ballard,

Hinton and Sejnowski, 1983; J.A.Feldman, 1985). Connectionist
architectures, by virtue of this parallelism, thus answer to the
demand that
> (cognitive) algorithms must be computable in amounts of
> time commensurate with human performance using the kind
> and amount of hardware that humans may reasonably be
> assumed to possess.(Rumelhart and McClelland,1985a, p.194.)

The second feature, robustness, amounts to a persistent functionality
under duress. It is a notable fact about human cognition that
although our memory of a given event, say, may become vague and
dulled with age or alcohol, it will seldom disappear neatly and
tidily in a 'here today, gone tomorrow' fashion. This gradual
tailing off is known as 'graceful degradation'. One explanation for
graceful degradation is to suppose that memories and representations
are distributed, with partial redundancy, across an entire network.
Connectionist models (as opposed to merely parallel ones) involve
just such distribution. The functional correlate of a specific
memory in a connectionist model is a total pattern of connectivity
amongst a whole system of units, none of which takes full
responsibility for encoding or retrieval. Retrieval depends on
prompting the system to restore an earlier pattern of activation.
This is possible due to changes in the weights of the connections
between units caused by the original experience. Such weightings
allow the overall pattern to be restored from a fragment acting as a
cue (see e.g. McClelland and Rumelhart, 1985 p.163). Because the
role of particular individual units in such a global process is not
decisive, the overall pattern of activation can often be restored
even after substantial damage to such units. Connectionist models
thus exhibit the graceful degradation characteristic of biological
systems. Perhaps their most interesting and potentially radical
features, however, have to do with the benefits accruing to them in
virtue of the form of representation they employ. It is these
features, to which we now turn, which encourage the radical claim
that connectionists are unearthing psychologically relevant facts
concerning the micro-structure of cognition.

III. RULES AND REPRESENTATIONS.

3.1 Superpositional representations. The most striking features of
(a large class of) connectionist systems concern the way in which
they embody rules and representations. To take representations
first, consider a connectionist model of memory. In such a model
(see McClelland and Rumelhart, 1985) memory traces will be stored
as alterations in the overall pattern of connectivity strengths
linking a set of units. In this sense the memory is distributed
across all of the units. Further 'experience' (the scare quotes
are there lest anyone object to crediting the system with too much
too soon) will alter that pattern but not so radically as to
obliterate all trace of the first experience. Subtle adjustments of
connectivity strength enable many traces to be simultaneously stored
in one pattern of connectivity. In this sense, the memory traces
are stored superpositionally. Such superpositional storage (in

which traces get superimposed rather than preserved in isolation)
enables the system to generalise its specific 'experience' in a very
natural way. The generalisation (which amounts to a strong
connectivity pattern corresponding to the shared features of specific
experiences) is an immediate consequence of the method of storage.
Nor does this necessitate loss of specific data about individual
exemplars to which the system has been repeatedly exposed. These can
be called up by cues consisting of a distinctive feature of the
exemplar in question. The connectionist model thus solves a number
of difficult problems simply in virtue of its method of storage.

The notion of superpositional storage raises important issues
concerning the extent to which it is appropriate to speak of
connectionist models as involving internal representations at all.
For stored representations, on the superpositional model, are
implicit and non-localised. Whereas conventional AI tends to
conceive of representations in terms of discrete, stored symbol
strings. The idea that multiple representations could be captured
in a single structure (viz. a connection ist network) with no
possibility of further localisation is, as Draper,1986 points out,
essentially alien to conventional AI. It is worth noticing that
such multiple functionality would be favoured by the demands of
economy which impose internal selective pressure on naturally
evolving systems (see e.g. Sober's,1981 development of the concept
of internal fitness).

3.2 Rules and constructivism. The notion of constructivism may be
traced back at least as far as the work of Emmanuel Kant. Human
thought, Kant suggested, depends on more than the fruits of
individual experience alone. It depends also on a lawful ordering
of experience imposed by the mind. In some sense or other, such a
claim (if not its Kantian transcendental idealist accompaniments) is
doubtless true. Individuals must be born with mental structures
suitable for the development of thought about the world, and this
will no doubt involve the projection onto the world of something
functionally equivalent to a set of expectations. This set of
expectations (or their functional equivalents) may also be augmented
and altered by subsequent experience to yield new and more powerful
'conceptual schemes'. But in what might such sets of expectations
consist? We may pursue the question by taking a brief look at
computer vision.

A recent text-book defines the vision problem like this;
         Given a two-dimensional image, infer the objects that
         produced it, including their shapes, positions, colours
         and sizes.
The process must involve inference, since;
         There is not enough constraint unless we make plausible
         assumptions about what the machine is looking at.
                         Charniak and McDermott, 1985, p.89,97.
The upshot is that vision systems need to embody some general
assumptions about the nature of the physical world. Such
assumptions might include; that texture gradients reflect surface

orientation, that rigid objects in motion are more common than
elastic objects deforming their shape, that surfaces are continuous,
and so on. Vision involves the satisfaction of a vast number of
such constraints. These need to be played off against each other.
This requirement of simultaneous satisfaction makes vision a prime
site for parallel, co-operative computational procedures. For if we
embody the constraints as differences in connectivity strengths in a
connectionist system, the process of relaxation described in II above
will yield a consistent solution to the problem (see Ballard, Hinton
and Sejnowski,1983 p.21). But notice what this means for the notion
of construction in a connectionist setting. Construction no longer
need involve the explicit encoding of inference rules or assumptions.
Such rules and assumptions are not represented in the system in the
explicit, localisable fashion associated with conventional AI.
Instead, functional equivalents of such rules and assumptions are
emergent from (i.e. implicit in) the structure of the network itself.

Potentially more radical still is the thought that connectionism, by
commanding a powerful model of learning, may substantially reduce the
amount of inbuilt (whether implicit or explicit) rules and
assumptions required by a would-be intelligent system. McClelland
and Rumelhart, summarising the nature of their model of distributed
memory, make the general comment that;
            Much of the appeal of distributed models is that they do
            not already have to be intelligent in order to learn.
                        McClelland and Rumelhart,1985 p.186.
An example of this in the field of vision may be found in Rumelhart
and Zipser,1985. This presents a connectionist model which, by a
process called competitive learning, develops a set of feature
detectors adapted to the set of inputs which it is given. The system
was able to learn to distinguish horizontal and vertical lines, and
thence to detect individual letters on a standard grid. It did so,
they stress, not as a result of some 'a priori set of categories
into which the patterns are to be classified'. Rather, it generated
its own feature detectors on the basis of the structure of the
stimuli to which it was exposed. The better structured the stimuli,
the better tuned were the detectors produced. In this way the
network can be seen to adapt itself to the domain in which it finds
itself. The model thus achieves the same kind of goal as the
Boltzmann machine described in Ackley, Hinton and Sejnowski, 1985
viz. to discover;
            Parallel organisations that do not require so much
            problem-dependent information to be built into the
            architecture of the network (and which) adapt a given
            structure of processors and communication paths to
            whatever problem (they are) faced with. Ackley, Hinton
            and Sejnowski, 1985 p. 148.
One clear benefit of connectionist models, then, is that they
enable AI to address developmental issues previously ignored or
glossed over (see e.g. Rutkowska,1984).

Connectionism, then, may presage a departure from strong
constructivism both by doing away with the need for explicit rules
and assumptions and by providing developmental models in which much

less initial structure is required. The biological benefits are
obvious. In an environment subject to occasional rapid change, and
given the slow and wasteful nature of adaptation by natural
selection, what could be more desirable than a _minimal_ amount of
pre-determined structure _maximally_ sensitive to whatever
regularities are encountered.

IV. IMPLEMENTATION MODELS AND PSYCHOLOGICAL THEORIES.

Connectionist approaches, we saw, embody some non-standard ideas
about rules and representations. The question therefore arises,
what is the status of such systems vis-a-vis their more conventional
cousins? Three main possibilities are suggested in the literature.
   (i)   Connectionism is just an implementation model of standard
          theories. It has no psychological relevance.
   (ii)   Connectionism displays the micro-structural detail of
          cognitive processes treated, at a lower degree of
          resolution, by conventional AI. This micro-structural
          detail is psychologically relevant.
   (iii) Connectionism explains a restricted range of mental but
          non-cognitive phenomena. Only conventional techniques
          can model truly cognitive effects.
In considering these options, and the sketchy evidence which exists
for them, I hope only to show (a) that the matter is quite unresolved
by the considerations adduced by the champions of (i) - (iii) and
(b) that the relation in question could turn out to be much more
complicated and interesting than anything listed above.

Option (i) is championed by D.Broadbent in his reply (Broadbent,
1985) to McClelland and Rumelhart,1985. He suggests that McClelland
and Rumelhart are wrong to present experimental psychological
evidence (see McClelland and Rumelhart,1985 p.173-183) as supporting
their claims about distributed memory. Such evidence cannot be
relevant, he claims, because the distinction between distributed and
non-distributed representations has no implications at the
computational level appropriate to psychology. _What_ function is
being computed, Broadbent notes, is a matter unaffected by the
details of the mode of computation. And psychology, he further
suggests, is concerned only with this computational level of
analysis. Whether the function is carried out by a simple Turing
machine or a massive parallel network is thus held to be
psychologically irrelevant. Connectionist models, to generalise
this argument, would be distinctive only at the level of the
implementation of some specified computational function.
Psychologically speaking such models could not be considered as
alternatives to, or deeper descriptions of, conventional approaches.
To so treat them would be to confuse levels of description.
Unsurprisingly McClelland and Rumelhart are not impressed.
Psychological relevance, they reply (1985a), extends below the level
of the theory of the computation (i.e. _what_ gets computed). It
extends to what Marr (1982) called the algorithmic or representational level which asks _how_ the function is computed. This involves
treating the way the information concerned is represented, the speed

of solution, the nature of performance under duress (e.g. noise) or
damage and so on. Any psychologically useful notion of computational
equivalence, they suggest, must be capable of distinguishing e.g. a
system which can compute a given function in real time from one
which cannot. On that kind of picture of computational equivalence,
connectionist models are not always computationally equivalent to
conventional models. A fortiori, they are not always mere
implementation models of the standard fare.

The response seems plausible. Psychology can hardly remain
indifferent to whether humans perform tasks in the manner of a
Turing machine or of a connectionist machine (but see my comments on
virtual machines in 5.3 for some reservations). The idea that
connectionist approaches, with their distinctive co-operative
algorithms and benefits of free generalisation and multiple storage
of representations, are necessarily doomed to psychological
irrelevance has, I think, little to recommend it. Option (i) may
thus safely be left aside. McClelland and Rumelhart's positive
claim, however, is less compelling. The claim (also in their 1985a)
is that - following our option (ii) - connectionist models describe
the (psychologically relevant) micro-structure of cognition.
Conventional models, on this picture, are a coarse-grained
approximation to the true micro-structural facts.

To clarify matters, McClelland and Rumelhart draw an analogy with
the relation of Newtonian mechanics to quantum field theory (op.cit.
p.196). Newtonian mechanics here corresponds to the 'macro-level'
theories of conventional AI. For some purposes, such a level of
description will do. At other times, however, a more fine-grained
account is positively required. At these times, the micro-level
theories (connectionist distributed models and quantum field
theory) must be called into play. But this analogy, I fear, is
ill-chosen. The picture emerging from it is one of conventional AI
as at best a handy approximation to the fine-grained truth studied
by the connectionists. Such a picture will almost certainly do
more harm than good, and it seems to me to underestimate the merits
of both conventional AI and connectionism. For, as we shall shortly
see, it is by no means obvious that connectionism and conventional
AI ought to study the same phenomena. It may be that they concern
different but inter-related aspects of human mental life and not
the same aspects at different levels of precision. If that were so,
then it could, of course, be the case that aspects of the mind
properly studied along connectionist lines have sometimes been
mistakenly studied by conventional techniques. In such cases, the
conventional models might indeed be at best a rough approximation
to the real thing. Indeed there may be conventional models of
phenomena (such as retrieval from memory) which are shown to be
entirely unnecessary and misguided by connectionist work. In this
way, connectionism might help to isolate the right areas in which
to apply conventional techniques. But it would not follow that all
conventional work was susceptible of so-called 'fine-grained'
connectionist rendering. Some aspects of our thought e.g.conscious
processes of step-by-step inference, have a phenomenology which

strongly suggests a more conventional model. If such a suggestion were borne out, the analogy with Newtonian mechanics and quantum field theory would have to be abandoned. For all physical phenomena are susceptible to quantum analysis. But there is as yet no proof that all mental phenomena are susceptible to connectionist analysis. Options (i) and (ii) thus depend on an unproven assumption to the effect that connectionism and conventional AI are, or should be, addressing the same range of mental phenomena. Option (iii) involves a particularly radical denial of this assumption and will occupy us in the next section.

V. SYMBOL MANIPULATION AND COGNITIVE EXPLANATION.

5.1 Symbols in a connectionist setting. Connectionism, according to Pylyshyn, 1984 p.69-74, 215, does not explain any truly cognitive achievements. At most, it explains the nature and range of primitive operations which define the virtual machine on which algorithms capturing psychologically relevant detail are run. The reason for this, if I understand Pylyshyn correctly, is that descriptions of connectionist work tend to depict the machines as finite state automata. And no machine so described can help explain truly cognitive capacities. It is important to stress that this argument turns not on what a connectionist machine could or couldn't ever achieve. Rather, it turns on how we must describe it in order to understand how it is capable of achieving whatever it does. Pylyshyn is quite explicit about this. 'What is at stake here', he writes, 'is the nature of the organisation captured in a certain description' (Pylyshyn,1984 p.72). To see what he has in mind, recall that a finite state automaton is a machine which is characterised by a finite set of state-transition sequences. These say that if the machine is in state S_1 and is given input Y it will go on to state S_2, if given input Z, it will go on to S_3 etc. etc. Such rules are often depicted by circle and arrow diagrams showing the possible transitions according to input. Chomsky long ago showed that human grammatical competence could not be captured by any such set of state-transition rules. This is because, very roughly, there is an arbitrarily expandable set of grammatical sentences having the same ending (since we can nest any number of clauses inside a particular sentence frame). Since any finite set of state transitions must include only a finite number of such completion equivalent grammatical sentences, it follows that grammatical competence cannot be captured by any such description. Instead, what we require is a finite set of rules applicable to an arbitrarily expandable set of symbol structures. This is a richer or higher level analysis of the system than that given in the finite state automaton description.

What Chomsky thus argued for grammatical competence, Pylyshyn urges for cognitive competences in general (Pylyshyn,1984 p.59-74). If he is right (an issue I shall not attempt to address here) then the explanation of cognitive phenomena is possible only via some high level analysis of a system into a control structure and some symbols upon which it operates. Finite state automata descriptions

fail to reflect such organisation and thus fail to explain cognitive
capacities. Supposing all this to be true, the following conclusion
about connectionist accounts would be warranted; that insofar as
connectionist work is described only in terms of state-transitions
in a connectionist network, it is of no use in explaining cognition.

Most current work in the connectionist tradition is so described
(see e.g.Hinton and Anderson, 1981), and to that extent the
criticism may be valid. But as a criticism of connectionist
explanation tout court, it would be premature, to say the least.
For it may be that there will be interesting higher level
descriptions of recognisably connectionist networks such that they
can be understood as manipulating symbols according to rules. We
are certainly in no position to rule this out for the very good
reason that no-one, Pylyshyn included, really knows what it means
for a system to 'operate on a symbol' anyway! One danger, to which
Pylyshyn may have succumbed, is that our ideas about what it means
to 'operate on a symbol' are still heavily influenced by
conventional AI implementations in which a symbol is a discrete
internal state, manipulable (copyable, moveable) by a processor.
One feature of connectionist machines is, we saw, their lack of
such discrete, easily manipulable internal states. But it is by no
means clear that these signal necessary conditions for a device to
be describable as manipulating symbols. Considered at a very
general level, a symbol is just a formal entity which is capable of
interacting with other entities in a way independent of, yet
systematically appropriate to, a semantic interpretation. So
conceived, it seems impossible to rule out some description of a
connectionist network which presents these functional properties of
systematic interaction. Any such description will no doubt refer to
patterns of activation in a network over time, and hence the symbol
correlates, whatever they are, will not be localised or fully
discrete. (One symbol may give way to another by a sequence of
small changes.) But as we said, these may not signal necessary
properties of machines described as symbol manipulators so much as
accidental properties associated with descriptions of conventional
implementations.

Is there perhaps a limit on what kinds of operations on symbols
connectionist networks could perform? Could a connectionist
machine, for example, compile a high level language? Once again,
we just don't know. But there seems to be no principled reason to
rule it out. Indeed, insofar as the basic structure of the brain
may conform to connectionist specifications, it could be held that
human beings already constitute an existence proof of some such
possibility. One way in which a connectionist machine might go
about compiling a high level language would be by learning in the
normal connectionist way. That is, it might be given examples of
inputs in the high level language and outputs in machine code. Or
it might be the case that some connectionist machine is able to
simulate a conventional architecture, in which case it would
(trivially) be capable of performing any computation expressed in a
high level language. (This process is also known as emulation and

and must be distinguished from mere simulations of input-output
profile alone see e.g. Pylyshyn,1984 p.98.) These two ways in which
a connectionist machine might be capable of complex symbol processing
(i.e. directly and by emulation) mark, I think, important
alternatives. For the psychological relevance of connectionist
models may ultimately depend on their being able to perform at least
some cognitive functions without simulating different architectures.
I discuss such matters further in the final section below.

5.2 Mixed relations and virtual machines. What, then, is the
relation between connectionism and conventional AI? The only honest
answer must be that it is too early to tell. One thing is certain,
however, and that is that options (i) to (iii) barely scratch the
surface of the possible relations which could hold between the two.
I should like to close by drawing attention to one source of possible
complications which - surprisingly - seems to have been largely
ignored in the discussions to date. It revolves around the
well-known notion of a virtual machine. Here is an argument put
forward by McCleiland and Rumelhart in their response to Broadbent.
        It is clear that different algorithms are more naturally
    implemented on different types of hardware, and therefore
    information about the implementation can inform our
    hypotheses at the algorithmic level.
                    McClelland and Rumelhart, 1985a p.193.
The implication seems to be that evidence to the effect that the
brain is a parallel distributed system counts in favour of an
approach to cognitive modelling which deploys co-operative
algorithms. But just what force does this kind of argument have?
The connectionist systems we currently study are after all themselves
(for the most part) simulated on conventional von Neumann
architectures. The reverse could conceivably be true of the brain!
Early biological pressures may have favoured the development of a
connectionist system as the real machine. But later cognitive
operations might depend on that machine's simulating a von Neumann
machine. If so, then the correct psychological theories of cognition
drawn up at the algorithmic level, would be those of conventional AI
and not those involving parallel distributed processing. It would
not even be correct to say that connectionism studied the micro-
structure of cognition, anymore than it would be correct to say that
conventional AI studies the micro-structure of those connectionist
systems which are simulated on von Neumann machines!

Let me end with some pure speculation. Suppose the human brain is a
connectionist machine which, at a late stage in its evolution,
'discovered' that for some purposes it was best to simulate a von
Neumann type architecture. For example, it might be that important,
basic and often repeated tasks are best dealt with by deploying
co-operative algorithms, but that where conscious ratiocination was
required, conventional algorithms run on a virtual von Neumann
machine were superior. If such conjecture were correct, then the
relation between connectionist models and conventional AI would be
non-uniform across the cognitive domain. Where the higher flights
of conscious ratiocination were concerned, the connectionist story

would indeed be irrelevant to an abstract psychological understanding
of those aspects of mind. In such cases, connectionism would be the
mere implementation detail which Broadbent believes it to be. But
where other tasks were concerned – tasks for which the brain did not
choose to simulate a von Neumann machine – the connectionist model
would be psychologically (i.e.algorithmically) relevant. In such
cases the connectionist would be straightforwardly studying the
structure (not the 'micro-structure') of cognition. It might even
be the case that all the so-called cognitively penetrable phenomena
turned out to involve the simulation of conventional architectures.
If so Pylyshyn's claims would be, to that extent, vindicated. (This,
presumably, is precisely the outcome he expects – see note 4 in
Pylyshyn, 1984 p.96.) Speculation aside, the moral is clear. There
is no a priori reason to expect a uniform relation between
connectionism and conventional AI across all cognitive domains. What
is sometimes just an implementation model may at other times
constitute a true psychological theory. The real task, if some such
complex relation held, would be to see how the various virtual
machines were organised into a single coherent system.

VI. CONCLUSIONS; PANACEAS AND PENICILLIN.

Connectionist models offer a powerful new tool to cognitive science.
The fringe benefits associated with such models (such as graceful
degradation, speed and free generalisation) are impressive and may
help avoid spurious attempts at constructing conventional theories
to account for such phenomena. Connectionism, moreover, provides a
new means for AI to address long-neglected developmental issues.
Concerning the vexed question of the relation between connectionist
models and conventional AI, it is too early to make definitive
pronouncements. But it seems unlikely that any simple relation (e.g.
'connectionism always displays the micro-structural detail of
standard theories'; 'connectionism is always just an implementation
model of standard theories') will prove to be correct. It may be
that the mind consists of many virtual machines with different
architectures, some connectionist, some more like von Neumann
machines. If so, then the important task is to see how these are
integrated into a single overall system and which tasks require
abstract descriptions in terms of which machines. Connectionism, on
that attractive picture, should provide a means of studying aspects
of cognition complementary to those studied in standard AI. If it
revolutionises cognitive science it should therefore do so in the
way penicillin revolutionised medicine; not by doing away with all
previous techniques, but by adding to them.

REFERENCES

Ackley, D., Hinton, G. and Sejnowski, T.,1985. A learning algorithm
    for Boltzmann machines . Cognitive Science 9,1985, p.147-169.
Anderson, A. and Hinton, G.,1981. Models of information processing
    in the brain, in Parallel models of associative memory (Eds.
    Hinton, G. and Anderson, A.) Erlbaum, New Jersey, 1981.

Ballard, D., Hinton, G. and Sejnowski, T. Parallel visual
    computation. <u>Nature</u> vol.306, Nov.1983 p.21–26.
Broadbent, D., 1985. A question of levels; comment on McClelland
    and Rumelhart. <u>Journal of Experimental Psychology: general</u>,
    vol.114 no.2 p.189–192.
Charniak, E. and McDermott, D.,1985. <u>Introduction to artificial
    intelligence</u> Addison–Wesley, Mass.1985.
Churchland, P.,1981. Eliminative materialism and the propositional
    attitudes, <u>Journal of Philosophy</u>, vol. LXXVIII, no.2.
Cummins, R.,1983. <u>The nature of psychological explanation</u>, MIT
    Press, Cambridge, Mass.
Draper, S.,1986. Does connectionism constitute a paradigm
    revolution? Draft paper – University of Sussex.
F eldman, J.A. Connectionist models and their applications;
    introduction. <u>Cognitive Science</u>, 9, 1985 p.1–2.
Gibson, J.,1979. <u>The ecological approach to visual perception</u>
    Houghton Mifflin, Boston.
Haugeland, J.,1981. Ed. <u>Mind design</u>, MIT Press, Cambridge, Mass.
Hinton, G. and Anderson, A.,1981. Eds. <u>Parallel models of associative
    memory</u>, Erlbaum, New Jersey, 1981.
Marr, 1982. <u>Vision</u> San Francisco; Freeman.
McClelland, J., and Rumelhart, D.,1985. Distributed memory and the
    representation of general and specific information <u>Journal of
    Experimental Psychology; general</u>, 1985 vol.114 no.2, p.159–188.
McClelland, J., and Rumelhart, D.,1985a. Levels indeed! A response
    to Broadbent <u>Journal of Experimental Psychology; general</u> 1985
    vol.114 no.2, p.193–197.
Michaels, C. and Carello, C.,1981. <u>Direct perception</u> Prentice Hall,
    New Jersey.
Pylyshyn, Z.,1984. <u>Computation and cognition</u> MIT Press, Cambridge,
    Mass.
Rumelhart, D. and Zipser, D., Feature discovery by competitive
    learning <u>Cognitive Science</u>, 9, 1 – 2,1985 p.75–112.
Rutkowska, J.,1984. Explaining infant perception; insights from
    artificial intelligence <u>Cognitive Studies Research Paper 005</u>,
    University of Sussex.
Sober, E.,1981. The evolution of rationality <u>Synthese</u>,46,1981.
Stich, S.,1983. <u>From folk psychology to cognitive science</u> MIT Press,
    Bradford.

INTELLIGENT MACHINES WHAT CHANCE ?

John Kelly

Department of Computer Science
University College Dublin
Belfield
Dublin 4
Ireland

Abstract

Forecasts of the likely development of intelligent machines are critically examined in the light of:

(a) present ignorance about human intelligence

(b) optimistic attribution of intellectual powers
    to computer programs

(c) ethical considerations

Contrast is made between the explicit nature of computer activities and the probability and importance of an ineffable component in the functioning of human intelligence. The increasing tendency on the part of some commentators to drop the use of scare quotes when referring to qualities like understanding, purpose, etc. is criticised. The nature of computers as for mal systems and the importance of the frame problem are emphasised.

The general conclusion is that it is still far too early to bring the concepts of machine and intelligence together in anything more than a primitively analogical fashion.

Keywords: machine intelligence, understanding, philosophy, ethics, formalism.

Introduction

In a recent speech at the *7th European Conference on Artificial Intelligence* Sir Clive Sinclair predicted that by the year 2020 machine intelligence would surpass human intelligence and that thereafter the intellectual functioning of machines would continue to expand with no obvious limits in sight. These predictions were made within the framework of a 'visionary perspective' but they were delivered in sober down-to-earth fashion and obviously represented the considered view of a highly successful practitioner of computer science. They echoed the words of the late Christopher Evans who went so far as to suggest that machines might eventually be "literally uninterested in us , conceiving human beings to be about as worth communicating with as we do hedgehogs or earwigs"(1).

The issues to be addressed in this paper are:

(a) whether and to what extent forecasts of this
     nature are justified,

(b) if justification is forthcoming whether forecasts
     of such extremity are palatable in terms of their implications
     for humans.

These questions have of course been subjected to prolonged critical examination by a number of writers in the past, most notably by Dreyfus(2) in the case of (a) and by Weizenbaum (3) in the case of (b). However, in view of their vigorous persistence and with regard to the currently high public profile of AI it seems timely to conduct a fresh examination.

In attempting to come to terms with an expectation that intelligent machines will be developed it is necessary to analyse the concept of intelligence and to determine whether it is accurate and fruitful to apply to machines a term which has traditionally been applied to animal life only. The question of whether animals, in particular humans are machines must be addressed. If humans are candidly considered to be machines then obviously a whole range of issues drops out of sight.

The complementary question of course is what is a machine ? This brings us face to face with the challenge of achieving a level of articulation of the concept of intelligence high enough to allow a principled description of the notion of intelligence to machines and opens up the possibility of identifying limits to such articulation. And what of the future ? Will we have truly intelligent machines by the year 2020 ? Or ever ? What ethical considerations arise ? At what stage, if ever, should our attitudes change from observation to normative, from tolerance to regulatory ?

Expectations

The view that machines might exhibit intelligent behaviour has a long history going back, at least, as far as Leibniz who dreamed of a general mechanization of reasoning and surely implicit in the strongly mechanistic view of the universe sanctioned by physics in the last century. However, the modern expectation was probably first clearly enunciated by Turing in his famous 1950 paper (4). "... I believe that at the end of the century the use of words and general educated opinion will have altered so much that one will be able to speak of machines thinking without expecting to be contradicted", and again "We may hope that machines will eventually compete with men in all purely intellectual fields".

To resolve any disputes that might arise concerning the success of that competition Turing proposed his celebrated Game Test. Although the test has been criticised over the years it still serves as a possible decision tool for issues of machine intelligence. Turing himself probably contributed to the flow of criticism by setting up some weak cockshot arguments against the validity of his test and rather too easily knocking them down.

The test gives rise to two major difficulties. The first one is the exaggerated expectation which emerges from the replacement of one of the human testees by a computer and the assumption that a suitable range of intelligent responses will be forthcoming. It does not help to dissolve the difficulty by saying that if the computer does not display acceptable behaviour it will have failed the test. The mere articulation of the computers' role sets up a feeling of near accomplishment. The naive observer is startled by any apparent manifestation of intelligence while even the experienced researcher is deceived into minimising the enormous range, subtlety, depth and difficulty of intelligence in action.

The second difficulty concerns the unashamed behaviourist nature of the test. Behaviourism (pace Skinner) has long ago been debunked as an adequate framework for understanding and explaining human behaviour (cf Chomsky 1959 (5), Koestler 1967 (6)). There is a distinct air of stimulus (question) - response (answer) in the description of the playing out of the game. As Weizenbaum has demonstrated it is all too easy to present a plausible picture of intelligence by means of a set of canned responses. And as other workers have realised it is the rapid descent into idiocy at the edges of competency that betrays the real lack of understanding and intelligence in the more serious simulations. We can easily imagine a very large store of delicately shaded canned responses to a wide variety of questions which might deceive even the most subtle and indefatigable interrogator. But surely this would be an elaborate conjuring trick. How would the computer decide for instance to respond at one time with a claim of ignorance and insensitivity and at another time with a few lines of heartbreaking sadness to the question of whether it wrote poetry or not. On the output of a random number generator ? And would this advance our understanding of what it is to have the gift of intelligence or move the machine beyond the limits of human intelligence ? Mimicry and deception are weak criteria. The fact that it is impossible to distinguish between photographs of a real Sir Clive and a very life-like statue hardly proves that the statue is alive. We would not ascribe chess mastery to a novice moving pieces over the board according to the directions of Gary Kasparov. *Mere* behaviour proves very little.

One of the fallacies which has continually fanned the flames of expectation is the assumption that we can build functioning artifacts without necessarily understanding the function to be achieved. "We must mention that there is one possible way of getting an artificial intelligence without having to understand it or solve the related philosophical problems. This is to make a computer simulation of natural selection in which intelligence evolves by mutating computer programs in a suitably demanding environment. This method has had no substantial success so far, perhaps due to inadequate models of the world and of the evolutionary process, but it might succeed". McCarthy & Hayes (7).

This ill-founded hope arose again at the **ECAI '86 Conference** when Sinclair proposed that the fact that aeroplanes were built without understanding how birds flew should serve as a basis for assuming that intelligence could be built without knowing what intelligence is. But this won't do.

Firstly, the building of planes did require understanding, not perhaps of the fine details of insect or bird flight but of some of the fundamental principles of aerodynamics. As understanding deepened, planes improved. When understanding was defective planes crashed.

Secondly, and this is a crucial point, the functioning of an aeroplane is entirely 'circumscribed' by physical parameters and reliance may be placed on the permanence and continuity of purely physical properties of materials. To assume that this is the case for intelligence is a serious begging of the question. No one now seriously believes that assembling a collection of "intelligent" elements in no matter how exact a replica of the human brain will generate intelligent activity in the absence of a deep understanding. Random mutations, etc., have had "no substantial success" and there seems no reason to believe they ever will. Expecting to create intelligent artefacts without understanding is a betrayal of intelligibility and of the whole methodology of science. This is not to say that building models in order to sharpen and deepen our understanding is to be eschewed but the confident prediction that somehow intelligence will emerge from the programming and simulation of isolated intelligent activities is surely not sanctioned by common sense or logic or the experience of science. To the extent that we rely on a happy coincidence of software and hardware and a re-enactment of the aeroplane phenomenon our predictions of ultra-intelligent machines are but a pious hope based on a mishmash of chance, probability and evolution.

It may well be argued that predictions of intelligent machines are not any more based on evolutionary processes determined by chance in even suitably constrained environments. This is probably true although such expectations are often subtly implicit in some work.

On what basis then are forecasts of intelligent machines made ? There are at least two:

(1) complex programs

(2) the forthright assertion that man is a machine

Some of the software running on today's machines is most impressive. Complex OS, DB systems, programs such as Dendral, Prospector, Macsyma can create a feeling of awe at their advanced problem-solving ability. This feeling is deepened on receipt of the information that probably no one person can fathom the complications of such vast programs. We are told that the statement that "A computer can do only what a programmer tells it to do" is true but misleading and we are asked to appreciate the reality of the programmer's surprise at the novelty of some of the programs paths and decisions.

But the blunt fact remains that the activity of the machine is completely determined by the programs and is in principle completely predictable. Only shortage of time or lack of energy or interest or the intelligent decision to let the machine get on with what it has been told to do at its own speed prevents a person from following through the implications of his code. Even so called "random" effects induced by random number generation are in principle determinable — after all they are the products of a program ! To the extent that the action of a machine is not determined by an explicit program it is accidental and therefore not intelligent. The functioning of the machine is completely dependent on the programs. It cannot go beyond what it is told in any realistic sense. Any departure from an explicitly described outcome has no significance for the machine - it causes no surprise, it is not understood. The autonomy of the human person stands in marked contrast to the dependence of the machine.

In spite of the present narrow, highly constrained and determinable behaviour of computers there is an increasing tendency among writers on AI and Cognitive Science to drop the alerting quote marks when ascribing qualities such as intelligence, creativity and purpose to machines. This has the effect of dulling one's critical awareness and of creating a temptation to read too much into the current capacities and operations of computers and has been described by Drew McDermott as "wishful mnemonics"(8). Thus for example Professor Margaret Boden, one of the most eminent, perspicuous and eloquent writers on AI exalts the blocks-world program SHRDLU in terms of its ability to *understand*. Quoting part of a dialogue conducted with SHRDLU she has this to say. "Items 17 and 23-29 show that SHRDLU has some understanding and memory of its own goal-subgoal structure, and is able to address this structure to find the reasons for which it did things"(9). And again in the same article: "In short these programs show the beginnings of one main criterion of purpose: variation of means". The phraseology is misleading Using computer or program as the subject of a sentence in this

forthright way lends an air of authority and conviction which is scarcely justified by the reality of performance. The programs are not really agents of
their own actions. Nor do they in any sense understand the reasons for their
actions. SHRDLU for example does not 'know' anything about real blocks.
It merely shuffles symbols around in a manner totally constrained by the
formalism of the program. It does not understand why a block cannot be
balanced on a pyramid. This rigid manoeuvring of symbols does not seem
to reach very far across the divide between machine operation and human
intelligence.

This gap, of course, is eliminated if humans are viewed as machines. One
doesn't need to be well versed in modal logic to see that machine intelligence
is possible if it actually exists as human intelligence in the actual world. This
view has been openly espoused by some scientists. Minsky refers to man as
a "meat machine" while the celebrated biologist Monod declares that "The
cell is a machine. The animal is a machine. Man is a machine"(10). If this
is accepted then, of course, the question of whether we will have intelligent
machines or not is solved. We already have intelligent machines — humans!
But before we agree to add humans to the set of machines we need to ask
what is man and what is machine. This issue will be addressed later on. In
the meantime it may be worth remarking that it is possible to **define** the
concept of machine so as to include man. But this may do violence to the
historical development of these concepts and our understanding of them and
may smother the important distinctions which bear on our apprehension of
reality. Even if some day an isomorphism is established between the human
brain and a machine it may be useful, nay it may be all important to retain
and emphasise the distinctions. Campbell has stigmatized the tendency to
identify machine functioning with human behaviour by labelling it the M-B
fallacy - 'Humans do B. Using M it is possible to do B. Therefore humans
must use M to do B' (11). The issue then is not easily resolved even by the
terminology of such eminent authorities as Minsky and Monod.

Intelligence and Machines

The computational model of human reasoning enjoys an eminent status
among cognitive scientists and contributes a powerful paradigm for theo-
rising, experimenting and philosophising. "The theoretical foundations of

cognitive science are rooted in logic and computation theory (Brainerd & Landweber, 1974; Minsky, 1967) from which Newell and Simon (1976) have derived their notion of 'physical symbol systems'. This notion defines a general class of systems that have the capacity to hold and transform symbols, or more generally, symbolic structures but exist as physical entities. Human beings and computers are the prime examples of such systems. (12) Again we note the disarming effect of the elevation of computers on to the same plane as man. Later in the same paper Slack declares that "Cognitive scientists believe that the notion of physical system is as fundamental to cognitive science as the theory of evolution is to all biology". (13)

In a sense the use of the phrase 'physical symbol system' brings us to the heart of the matter. Leaving aside for a moment the question of whether human beings can be comprehensively encapsulated by such an economic phrase it is clear that it is an apt pointer to the identity of a computer. A computer is essentially a formal system, a collection of symbols. This becomes clearer if we reduce the emphasis on the actual hardware which in fact contributes nothing fundamental to the meaning of a system, and concentrate instead on the reality of software. A program is a set of strings in a formal system — "A logic program is a finite set of program clauses" (14) where program clauses are a restricted form of well formed formulae of a certain type. Somehow we are less likely to attribute intelligence, whatever it may really be, to even the most well organised inscriptions on a sheet of paper. A specious appearance of some sort of ascent into autonomy, some sundering of the fetters of formalism is created by the dynamic activity of a physical electronic embodiment of formal systems.

This impression is strengthened, as we have already noted, by the continuous (and it must be admitted sometimes inescapable) use of the computer as the subject of sentences. "A machine is intelligent if it solves certain classes of problems requiring intelligence in humans or survives in an intellectually demanding environment"(15). But computers don't solve problems except in a purely analogical manner. A machine doesn't solve say sorting problems any more than a plane solves the problem of transport. Humans solve problems using computers or planes or sometimes even other humans. In fact machines don't have problems. There may be problems with machines bu not for machines. Take the sorting problem. A machine has no concept of sorting, has no idea of the intention behind the request to arrange, for example, in alphabetical order. It doesn't know:

- that the activity is based on agreed conventions

- that there are numerous other arrangements possible

- what the difference is between ascending and descending

- that sorting makes searching more efficient

- what it means to be more efficient

- what it means to search

- that being last on a list can offend people

- that being first on a list can improve a candidate's chances in an election

- that there is a christian adage about the first being last and the last being first, etc.,etc.

The machine doesn't muse on different aspects of the result, doesn't adopt different points of view, doesn't adopt any point of view. Just as well, it may be argued, since considerations such as the above are mostly irrelevant. But there's the rub. The machine doesn't know that. It has no concept of relevance. And this gives rise to perhaps the most serious problem of all for AI - the frame problem. What aspects of a problem situation or context are relevant ? How are the important features selected and the 'infinite' rest ignored ? How are new considerations brought into view in a selective and economic way as required ? So central is this problem for the evaluation of intelligent activity that we might partially characterize intelligence as:

- the ability to (partially) solve the frame problem.

The apparent manifestations of intelligent problem solving in advanced AI software, even in 'creative' programs such as DENDRAL and AM, are exposed as inadequate by contrast with the frame problem. Intelligence is not simply a collection of even finely honed 'intellectual' tools. It is rather too naively reductionist to expect that the nature of human intelligence will be revealed by developing a battery of tools for solving difficult problems. It must be acknowledged also that a problem which requires the application of advanced intelligence on the part of a human such as playing a master game of chess requires no intelligence on the part of a machine since the machine is simply blindly obeying the rules of a formal system. Well might the machine reply to a question regarding the motivation for a felicitous move 'don't ask me, I only obey orders'! Although computer programs are intensely teleological their purposes are due entirely to a human programmer. Before the modern computer came on the scene the concepts of machine and intelligence had been opposed to each other. Indeed when we speak of a person acting like a machine we mean he is acting *without intelligence.* The ability of computers to carry out enormously complex computations and to accomodate the spice of pseudo random injections have clouded the traditional distinctions. The inexorable pursuit of the logical implications of a formal system certainly gives the impression of great intelligence at work. After all only highly intelligent humans with special mathematical gifts and training are able to carry out this work. But again we must insist that while intelligence is required of humans it is not required in machines. The sophistication of computers at work on AI problems does not absolve us of the charge of doing violence to the traditional notion of machines and intelligence by glibly associating the two concepts. This can be seen to be the case even in the absence of complete definitions of machine and intelligence. The matter hinges on the usefulness, historical propriety and intellectual integrity of retaining a distinction between the operations of machines and the operations of human intelligence.

We have referred to the deep distinction between humans and machines in their ability to solve the frame problem. Related to this is another characteristic of human intelligence - the ability to cut the Gordian knot of circular argument or definition to seize the essential meaning of a concept. Consider the tangle of inter-referential words used in any dictionary to define a notion such as *being,* for example: *being - existence - exist - have place as a part of objective reality, etc..* No mere formal transformations of symbols can build the coherent yet flexible understanding of the concept of being that most humans have. The apprehension of meaning appears to have an ineffable aspect which may well be essential for the proper functioning of intelligence. This has been comprehensively explored by Polanyi in his beautiful work 'Personal Knowledge' (16). The possibility of a tacit dimension to knowledge points to the probability of inescapable limits to the development of 'formalised intelligence' and exposes as altogether too cavalier the approach of McCarthy & Hayes in their casual observation that "It is not difficult to give sufficient conditions for general intelligence. Turings' idea that the machine should successfully pretend to a sophisticated observer to be a human being for half an hour will do"(17). It casts doubts on the achievability of the desires of the same two scientists. "We want a computer program that decides what to do by inferring in a formal language that a certain strategy will achieve its assigned goal. This requires formalizing concepts of causality, ability and knowledge"(18).

The gap between reality itself and our representation of it is a serious problem. We live in a continuous environment but we discretise it for the purpose of articulation. Something may be lost on the way. This loss is often increased when the elements of articulation are frozen in a formal system. For humans symbols and words are not absolute entities standing apart from the apprehension and understanding of reality. *They are part of that understanding* and have the effect of triggering a rich resonance of meaning for other humans. But not for machines. There appears to be a certain poverty in mere articulation where the contribution from the tacit dimension is missing.

The shifting role of symbolism creates a curious tension between man and machine. For humans, words are arbitrary in nature but connected to reality in a far from arbitrary fashion while for machines words have a non-arbitrary character but are not connected to reality at all. The machine gets very upset if, for example, 'while' is replaced by 'as long as' but evinces no sign of disconcertion when it computes a new total of -1 for the size of the earths' population !

None of this is to assert that man is a prisoner of the connection between language as is and reality, nor to deny the powerful productivity immanent in the use of abstract symbols. But formalism has its limits as Gödel and others have shown. And along the tacit dimension there appear to be realities that are simply not sayable.

Perhaps at the root of the reification of intelligence in machines is a belief in the infallibility of the scientific method. The danger of this belief is the seduction of its apparent well-foundedness and the difficulty of adopting a meaningful critical point of view. After all the very tools of criticism are among those of science itself - analysis, logic, experiment etc. But the warning signs are there. Kuhn has revealed the role that fashion, social emphases, economic factors, and other non-objective influences play in determining the direction that science takes (19). The excesses of weaponry and war underline the horror of science out of control. The fetishism of efficiency applied to the production of food mountains in contrast to the problem of mass starvation exposes the poverty of scientific logic.

Future

The last remarks may seem out of place in a paper attempting to analyze the grounds for the possibility of machine intelligence. But ethical considerations cannot be ignored when we contemplate the likelihood of a serious rival to mans' intellectual hegemony.

The forthright view of man as machine already alluded to raises some awkward moral questions. On what basis other than utilitarianism could one argue against; apartheid, racism, genocide, economic tyranny, and so on. Can we afford to eschew a deep reverence for human existence which draws upon a radical distinction between man and machine ? The arguments of those such as Margaret Boden who would humanise the machines are less than convincing. The pass would appear to be already sold by the declaration "I believe a basically mechanistic view of the universe, and of human beings as creatures in it, to be true"(20). The attempt to reduce human knowledge, purpose, intelligence, responsibility to mechanistic primitives gives a philosophical underpinning to those who already profit from the 'thingifying' of humans. And of course, the temptation is great for all of us. Philosophically, it is easy to succumb to the tyranny of the obvious and pragmatically to yield to the exigencies of social organisation. Mechanical man walks, talks and we think, thinks before our eyes. It is convenient to contemplate units of production, consumers, market sizes, dole queues. But is that all ?

Let us imagine a world populated by ultra intelligent machines as forecast for us by the seers of a computer universe. What will the machines do ? Solve problems perhaps. But what problems ? Deprivation and starvation, economic imbalance, threat of nuclear holocaust, ecological ignorance, how to get on with your neighbour, how to keep sane, what ? The problem is that these are not problems for the machines. These are *human* problems. They relate to *human concerns*. Perhaps, contrary to the projections of some extremists, human life will be preserved by the machines and they will spend their time solving human problems. But why ? What price their superior intelligence ? Perhaps they will conduct a far reaching scientific exploration of nature and unravel the riddles of the universe. To what end, for what purpose ? Will we pollute them with our human purposes ? Or perhaps the machines will spend all their time building more and more intelligent machines down the lonely reaches of a humanless infinity.

Not many engage in such far reaching projections or confront the deep ethical and existential problems which arise. But there is a more immediate problem posed by the too ready conflation of human and machine intelligence. That is the danger of a redefinition of intelligence. This accusation has usually been thrown at those who try to wriggle out of the difficulty caused by chess playing, theorem proving, diagnosing, hypothesis generating programs by proposing other intelligent tasks that humans can do but that computers currently can't. As we have seen this is a naive view of the strength and implications of those programs. The real danger is the other way. Intelligence may be subtly redefined as the functioning of systems that 'solve' a collection of suitably difficult tasks. The problem is these tasks are no longer hard when solved by an algorithm. Intelligence is not required to carry out the prescriptions of a master level chess program in the same way as it is needed to cross a busy road successfully. There is a serious danger of narrowing the scope of the intellectual universe by a wish-fulfilling ill-founded optimism about machine intelligence.

So what is a reasonable attitude ?

A desire for and expectance of creative symbiosis !

Rather than elevate machines to the level of intellectual gods let us more modestly and with better hope of success emphasise the prosthetic potential of computers. Let us return man to the centre of the universe. Machines Machines can carry out our intellectual bidding faster, more accurately, more tirelessly than we have ever dreamed possible. We must learn better how to tell them what to do.

Summary

The discussion in this paper has been concerned with justifications for the forecasts of machine intelligence. The argument is not that machine intelli-

gence is impossible but that developments to date do not support strongly
optimistic projections. The current practice of using "wishful mnemonics"
and the view of man as machine tend to distort the analysis of the problems
and undermine the need to make and explore important distinctions. Any
possibility of real intelligence fundamentally different from mans' is as yet
a chimerical prejudice or represents an impoverishing change of definition.
It is conditioned by a longstanding preference for certain forms of scientific
knowledge and rationality to the exclusion of other possibilities.

There is still a wide (perhaps unbridgeable) gap between the complexities
of formal systems even when dynamically implemented and the richness of
human intelligence. The ineffable tacit dimension of human intelligence may
well prove to constitute a fundamental barrier to a reduction to atomicity
and a subsequent synthesis of composite intellectual activity. In addition to
engineering advances a great deal of research into the philosophy of mind
and machine is required before 'machine intelligence' becomes a principled
phrase in relation to future possibilities.

NOTES AND REFERENCES

1. **Evans** , Christopher. 1979. **The Mighty Micro**, Cornet books. Hodder and Stoughton. 242.

2. **Dreyfus** , Hubert L. 1979. What Computers Can't Do. **The Limits of Artificial Intelligence**. 2nd Edition. Harper & Row.

3. **Weizenbaum** , J. 1984. **Computer Power and Human Reason**. Penguin London/Freeman, San Francisco.

4. **Turing** , A.M. 1950. Computing Machines and Intelligence in **Computers and Thought**. (Eds Feigenbaum & Feldman, Robert E). Krieger Publishing Co. 1981.

5. **Chomsky** , N. 1959. Review of B.F. Skinner's verbal behaviour in **The Structure of Language**. (Eds Fodor & Katz). Prentice Hall, New York. 1964.

6. **Koestler** , Arthur. 1967. **The Ghost in the Machine**. Pan Books.

7. **McCarthy** , J & Hayes, P. 1969. Some Philosophical Problems from the Standpoint of Artificial Intelligence in **Machine Intelligence**. 4, 463-502, Edinburgh University Press.

8. **McDermott** , Drew. 1976. "Artificial Intelligence meets Natural Stupidity " in **SIGART Newsletter** 57.

9. **Boden** , M.A. 1984. AI and human freedom in **Artificial Intelligence-human effects**. (Eds. Yazdani & Narayanan). Ellis Horwood, London.

10. **Monod** , J. 1972. **Chance and Necessity**, Collins, London.

11. **Campbell** , J.A. 1984. Three Uncertainties of AI in **Artificial Intelligence - human effects**. (Eds. Yazdani & Narayanan). Ellis Horwood, London.

12. **Slack** , Jon, M. 1984. Cognitive Science Research in **Artificial Intelligence - Tools, Techniques and Applications**. (Eds. O'Shea & Eisenstadt). Harper & Row.

13. **Slack** , Jon, M. 1984. Cognitive Science
Research in **Artificial Intelligence - Tools, Techniques and
Applications**. (Eds. O'Shea & Eisenstadt). Harper & Row.

14. **Lloyd** , J.W. 1984. **Foundations of Logic Programming**.
Springer-Verlag.

15. **McCarthy** , J & Hayes, P. 1969. Some Philosophical
Problems from the Standpoint of Artificial Intelligence in
Machine Intelligence. 4, 463-502. Edinburgh University Press.

16. **Polanyi** , M. 1962. **Personal Knowledge**. University of
Chicago Press.

17. **McCarthy** , J & Hayes, P. 1969. Some Philosophical
Problems from the Standpoint of Artificial Intelligence in
Machine Intelligence. 4, 463-502. Edinburgh University Press.

18. **McCarthy** , J & Hayes, P. 1969. Some Philosophical
Problems from the Standpoint of Artificial Intelligence in
Machine Intelligence. 4, 463-502. Edinburgh University press.

19. **Kuhn** , T. 1962. **The Structure of Scientific Resolutions**.
The University of Chicago Press.

20. **Boden** , M.A. 1984. AI and human freedom
in **Artificial Intelligence - human effects**. (Eds. Yazdani
& Narayanan). Ellis Horwood, London.

Representing Time

THE LEXICON, GRAMMATICAL CATEGORIES AND TEMPORAL REASONING

Alexander Nakhimovsky

Department of Computer Science
Colgate University
Hamilton, New York 13346

1. INTRODUCTION

This paper proceeds from the premise that a natural language understanding system has to have a great amount of commonsense knowledge indexed by lexical items. It concentrates on one cluster of commonsense knowledge -- time and temporal reasoning -- and asks the following questions: What part of this knowledge belongs in word meanings? How is it represented there? How does it interact with the temporal reasoner? To illustrate the knowledge we are after, consider the following simple dialogs and narratives:

1. A. Can we meet for lunch today?
   B. I have a 12:30 class.

2. The glass ball was falling towards the table. Harry tried to catch it but failed. We spent the next hour picking up the pieces.

3. (a) The boy built a boat quickly.
   (b) The boy was building a boat quickly.

4. (a) After supper, Alice and Sharon sat down to read. Alice read her book quickly, turning the page every couple of minutes; Sharon read slowly, moving her lips as she made her way down the page.
   (b) After supper Alice and Sharon sat down to read. Alice read her book quickly and went around the house looking for something else to do.

In (1), in order to understand that B's answer is no, we need to know the time and duration of a prototypical lunch. In order to understand (2), we need to know not only approximate durations of events but also their internal constituency. The event of a ball falling on the table consists of three stages: the ball begins to fall, the ball is falling, the fall is coming to an end with the event of the ball hitting the table. The first sentence of (2) puts us in the middle of the falling process; the last sentence of (2) puts us in time after the falling event; we infer that the process came to its natural end. A useful analogy is that a sentence describing a process puts us in the middle of a script that includes

the beginning and the end, or several possible ends, of that process. I will say that a process, its beginning and its end stand in
compositional relations to each other.

The contrast between (3a) and (3b) shows that the meaning of the
word quickly depends on whether it modifies the description of a
process or an event: the meanings are 'at a faster than usual rate'
and 'in a shorter than usual duration,' respectively. (4a) and (4b)
show that in a language like English, which lacks a perfective-imperfective aspectual opposition, the same sentence Alice read her
book quickly can describe either a process or an event. The property of being an event or process cannot thus be associated with a
class of lexical meanings or with a specific verb form (progressive). Rather, it has to do with the movement of narrative, which,
in turn, depends on shifts of focus, the 'temporal schema' or 'aspectual class' of the described history, and the tense/aspect system
of the language.

The three kinds of knowledge that examples (1)-(4) illustrate (durational, compositional and aspectual) are the main subject of this
paper. Section 2 discusses durations. Section 3 clarifies the
relation between grammatical aspect and the notions of process and
event. Section 4 develops an aspectual classification of lexical
meanings, and section 5 presents an extended example and points out
possible directions to pursue.

2. DURATIONS AND SCALES

The main class of objects this paper is concerned with consists of
situations evolving in time, best visualized as four dimensional
pieces of time-space incompletely specified by their situation
types. Loosely following Hayes 1978, I speak of history-tokens
(h-tokens) corresponding to Hayes' histories, and history types
(h-types), corresponding to situation types of Barwise and Perry. I
assume that associated with each predicate of the meaning representation language (RL) is a set of roles such as agent, object or
patient; an h-type is a predicate together with its roles and a
selectional restriction on them, which is usually a conjunction of
sortal specifications for each role of the predicate (Hobbs et al.
1986).

The function **time-of** takes an h-token and returns its **lifetime,**
which is an interval or set of intervals. I thus adopt the position
that temporal logic should be event based (van Benthem 1983) and
that temporal relations between h-tokens are not reducible to the
global ordering on the time line. Included in these relations is
meet or **immediately precede:** the beginning of one interval can be
directly after the end of another one, with the restriction that
events can only meet processes or states, while processes and states
can only meet events. (The obvious idea is then to associate events
with closed intervals, states and processes with open ones.) The
intuition behind the proposed restriction is that events are discrete countable entities with a distinct beginning and end; packaged
in between the beginning and end of an event is the "stuff the event

is made of", which is a process or state. This intuition is discussed in a considerable body of literature that compares the event
-process and count-mass oppositions (Mourelatos 1981, Bunt 1985,
Bach 1986). As I argue below, all these authors should also have
allowed for events made out of states, as, for example, the event
described by Bobby took a nap.

In order to capture the intuition that every event token (e-token)
has a distinct beginning and end, I assume two functions, **begin-of**
and **end-of**, from e-tokens to e-tokens. It seems reasonable to assume that the duration of an e-token is by at least a constant factor greater than the durations of its beginning and end. (I make
this more precise below.) Since **begin-of** can be applied to the
yield of **begin-of** there are two possibilities: either we obtain a
new e-token and a new duration every time we do that, in which case
intervals can become infinitely small; or at some point we obtain an
indivisible interval, equal to its beginning and end. Data from
both linguistics and psychology suggest that there is a threshold
below which we cease to perceive an event as consisting of a beginning, a middle and an end; languages select such events for special
treatment by making it impossible to place the discourse time inside
them. (So, for example, the sentence The light was flashing describes a sequence of flashes, because the progressive of flash does
not put the discourse time in the middle of an individual flash.)
The following definition is thus justified: an e-token is instantaneous iff it is equal to its beginning and end.

On all non-instantaneous e-tokens the function **body-of** is defined
whose value is the process or state that the argument e-token 'consists of'. Unlike instantaneous events, which are all of the same
'instant' duration, non-instantaneous events have different durations, which can be compared and measured, with cyclical events providing units of measurement. Some of the units are determined by
celestial and biological cycles, others are civilization specific.
It is commonly assumed in AI literature that the choice of a measurement unit is arbitrary and independent of the interval to be
measured, because we can always switch from one unit to another by
using axioms of the sort (=(hour (* 3600 second))). Mathematically,
of course, this is true, but I don't think language quite works this
way: when we say It took me an hour to repair the faucet, we don't
mean It took me 3600 seconds to repair the faucet. Given an interval, we choose a unit to measure it with so that the duration is
expressed by a 'reasonable' number, that is not too big. Small
fractions are also avoided: we say half a month or a quarter of an
hour, but not one fifth, sixth and so on. (Note that a quarter is
the smallest fraction for which there is a special word; the names
for the rest of them are formed by a regular morphological process.)
Under these, psychologically very understandable, restrictions it is
rarely the case that an interval fits precisely into the measurement
unit we use, which results in a certain degree of fuzziness: when we
say that something happened a year ago we typically don't claim to
be precise to the day. The same imprecision is observed when we use
man-made time units such as hours, minutes, and seconds. The sentences Joe slept for three hours, or Joe slept from 2 to 5 do not

claim precision to the minute; in fact, the second sentence can be
continued with He woke up at a quarter after 5 feeling greatly re-
freshed, and the narrative does not become self-contradictory.
Formally, then, time intervals have durations of two kinds: precise
durations which are real numbers, and 'commonsense' durations which
are fuzzy numbers. There are, correspondingly, two functions, **dur**
and **dura**, that relate intervals to durations. (All functions that
have interval arguments are also allowed to take h-tokens, to which
time-of is automatically applied to obtain intervals.) **Dura** takes
an interval and a unit of measurement and returns a real number.
Every interval has units of measurement that are 'natural' for it:
measured (by **dura**) in a natural unit, the length of the interval
will not be a very small fraction (greater than some constant R) or
a very big number (less than some constant N). The function **scale**
-of takes an interval and returns the sequence of its natural units,
a subsequence of all the duration units.   The other duration function, **dur**, takes an interval and returns its duration, as a fuzzy
number, measured in the largest of the interval's natural units.
Examples of fuzzy durations are: [1 1 year] 'a year, about a year';
[1.5 1.5 week] 'a week and a half'; [2 3 day] 'two-three days' and
so forth. There are rules that transform fuzzy durations in one unit
into fuzzy durations in a smaller unit, but only as long as both
units are natural to the interval being measured: [1.5 1.5 week]
becomes [9 12 day], but not [200 260 hour].

The fact that hours are simply irrelevant when we talk about weeks,
or days when we talk about years, is expressed by the relation **much**
greater than (>>), defined as follows: i >> j iff the largest natural unit of j is smaller than the smallest natural unit for i.
There are thus two ordering relations defined on durations measured
in the same scale: **greater-than** (>) and **much-greater-than** (>>);
together, they give rise to the **equi** relation: two intervals' durations are **equi** iff the smaller of the two durations is much greater
than the difference between them.   The **equi** (indistinguishability)
relation is thus derived, rather than primitive as in Hobbs 1985,
Hobbs 1986, and it is not incompatible with **greater than**. I find it
both convenient and intuitively correct that we can treat two intervals as equivalent without losing the knowledge, if we have it, that
one is greater than the other. From the properties of the two ordering relations it can be shown that the **equi** relation is fuzzy,
i.e. symmetric and reflexive but not transitive.

The >> and **equi** relations constrain the relative durations of the
components of a non-instantaneous e-token: the e-token's body is
much longer than its beginning and end. This axiom is weaker than
the requirement that begin and end events be instantaneous, but it
is probably more accurate: note that the implied duration of He's
finishing his work may range from minutes to months depending on the
duration of the work in question. Although relatively weak, the
axiom supports the consequence that the duration of a non-instantaneous e-token is **equi** to the duration of its body.

Durations and time scales of h-tokens are constrained by their h-
types, i.e. lexical meanings; the important principle is that every

lexical entry describing a history ought to have durational information, much of it inherited via ISA links. For some h-types, the
duration of their h-tokens is stable and can be entered in the lexicon directly; for others, the scale is quite narrowly constrained;
and for most h-types, the duration of their h-tokens monotonically
depends on certain parameters of their participants. (The time required to read a book depends on its size and genre, and on the pro-
ficiency of the reader.) Information about durations can thus be
entered in the lexicon in the following three ways, which are not
mutually exclusive: (a) directly as a fuzzy duration; (b) as a
scale, i.e. a sequence of measurement units that are natural for the
h-type; (c) as a function of the participants of the h-type and
their various parameters. (Functional dependencies are further
discussed in section 5.)

Specifying durations and scales in terms of measurement units pre-
supposes that the units themselves are 'defined,' i.e. mutually constrained by a set of axioms. The required axiomatization of the
clock and the calendar involves identifying all the time measurement
units, defining them in terms of basic cyclical events, and es-
tablishing the >> relation on them. Once this is done, the lexical
entries for words denoting durations can be written. Since the
measurement units are determined by astronomy, biology and history,
the same word may have meanings corresponding to two or three of
these spheres of knowledge. Day, for example, in addition to its
'scientific' definition as a 24-hour interval between two consecutive midnights, has three 'commonsense' ones, based in nature, human
biology and social life, respectively: a stretch of time of varying
length between a sunrise and the next sunset, a stretch of time between waking up and going to sleep, and the 9 to 5 work day. The
scientific definition is probably the least commonly used, especially as part of the definitions of tomorrow and other indexicals: the
sentence I'll call her tomorrow means 'I'll do it between the time I
wake up and the time I go to sleep' or 'I'll do it during the next
work day.' A lexicological analysis of time measurement units will
probably yield many examples of the geo- and anthropo-centrism per-
vading language and commonsense reasoning.

3. ASPECTS AND EVENTS

This section establishes major divisions within a classification of
lexical meanings based on their time-related properties. Such classifications interact in complex ways with the time-related grammatical categories of the verb, and, in fact, what is grammatical in one
language may be lexical in another, with a range of disputable cases
in between. In English, for example, progressive and non-progres-
sive are clearly grammatical forms of the same lexical item, while
in other languages a given verb has a constant aspect value that
strongly constrains (although never fully determines) the temporal
properties of h-tokens that a sentence with this verb can describe.

The non-progressive--progressive aspectual system is the less common
of the two main 'varieties of such systems. The more common variety
is the perfective-imperfective opposition, which (Comrie 1976:4)

defines as follows: "perfective looks at the situation from the outside, without necessarily distinguishing any of the internal structure of the situation, whereas imperfective looks at the situation
from the inside." To formalize Comrie's definition I introduce the
concept of the Active Window on the Discourse, or Window for short.
Window can be thought of as an open interval of 'instantaneous'
length that, in a narrative, provides an answer to the question
"Where in time is my narrative now?" A sentence with a perfective
verb locates the Window immediately after the history that the sentence describes; a sentence with an imperfective verb locates the
Window inside the history.

Next to the perfective and imperfective aspect of the verb, it is
important to contrast the perfective and imperfective view of a history expressed by a sentence: the imperfective view puts the Window
inside the history, the perfective view puts the Window after it.
All languages, irrespective of their aspectual system, need this
contrast and are capable of expressing it. Perfective sentences
describe sequences of events and move the narrative along; imperfective sentences describe simultaneity, providing background, descriptions and such (Hopper 1978:216). In what follows, it is essential
to keep the following sets of concepts apart: perfective and imperfective aspects of the verb; perfective and imperfective view on an
h-token; the intrinsic properties of histories (h-types) which make
their tokens more or less appropriate to be viewed perfectively or
imperfectively.

The aspectual system of English is such that it has a variety of
imperfective (the progressive aspect) but no perfective, i.e. no
verb forms that unambiguously signal, in and of themselves, the
perfective view on an h-token that could also be viewed imperfectively. The last clause is important because instantaneous events
are intrinsically incapable of being viewed imperfectively: the
Window wouldn't fit. It doesn't follow that the corresponding verbs
are 'perfective' because aspect categorizes linguistic expressions,
not histories. (So, flash is not a perfective verb even though the
sentence A bolt of lightning flashed unambiguously expresses the
perfective view.) For verbs describing non-instantaneous h-types,
the same verb (e.g. read), in the same simple past form can, as we
saw, assume either aspect. (Recall examples 4.) However, in zero
context the sentence Alice read her book quickly is interpreted as
"perfective" because the history it describes has a natural end to
it: Alice finished the book. Just as in the case of instantaneous
events, we wouldn't want to say that the corresponding verb is "perfective" because the perfective view is not signaled by the form of
the verb but rather inferred, both from the form of the verb (non-
-progressive) and the following intrinsic property of the history:
whether or not the history has a built-in terminal point, which is
reached in the normal course of events, and beyond which the history
cannot continue. Following Comrie 1976 I use the terms telic-atelic
for the distinction.

An important point is that the oppositions telic-atelic and perfective-imperfective are independent of each other. In languages with

a well articulated perfective aspect it is quite obvious that a
stretch of a most atelic activity can be viewed perfectively, i.e.
as an event. English, lacking a grammatical category, employs a
variety of lexical means to achieve the same effect. A good example
is He took a nap, which indicates a perfective view, even though the
body of the described h-token is not telic, and not even a dynamic
process. Surprisingly, collocations of this kind have never, to my
knowledge, been discussed in connection with the English aspectual
system. Other possible constructions are do some V-ing and go
V-ing, as in I went skiing twice yesterday and did some reading in
between.

Atelic histories, of course, lend themselves more easily to the imperfective view and telic histories to the perfective one. Telicity
and perfectivity come together in, perhaps, the most common inference at work in narrative understanding: given a telic history
viewed perfectively assume that its end-point is reached. (See
section 4.4 for a more precise statement.) For all that, they remain distinct, if not entirely orthogonal, categories. We thus
have a distinction between h-tokens, grammaticalized in some languages, which correlates rather well with a distinction between
h-types, or lexical meanings; in languages where the token level
distinction is not grammaticalized, the tendency is to lump it together with the type level distinction. So, many aspectual classifications of English verbs lump telicity and perfectivity together
under the name accomplishment (Vendler 1967) or event (Mourelatos
1981, Allen 1984). However, from the typological linguistic perspective the terms 'event' and 'process' are simply another way of
talking about aspects: "...the term 'process' means a dynamic situation viewed imperfectively, and the term 'event' means a dynamic
situation viewed perfectively" (Comrie 1976:51). Note that if the
definitive property of events is that they can be counted (as in
Mourelatos 1981), then certainly there are atelic events: you can
take a nap or go skiing twice, although the histories are atelic.

The foregoing has established that telic-atelic is a major division
within dynamic histories or processes. It is generally agreed that
processes ought to be distinguished from static histories or states.
Dahl 1985:28 observes that static-dynamic is the "most salient opposition" within his "taxonomy of situations." As I hope to have
shown, both states and processes can be viewed either internally, as
h-tokens with open time intervals, or externally as events. Both
states and processes are thus opposed to instantaneous h-tokens that
can only be viewed externally as events.

4. A DETAILED CLASSIFICATION

4.1 States. States are further subclassified according to whether
being in a state is subject to volitional control. The English progressive is sensitive to this feature, as can be seen most clearly
in sentences with nominal predicates: while She is being funny is
normal English, The sea is being choppy is not. (Vendler
1967:102-3) defines state terms as those verbs that lack continuous
tenses and combine with durative time adverbials. He then mentions

the absence of volitional control "as a surprising feature about
states" which, he thinks, "is not strictly connected with considerations of time" (p. 104). Most scholars use the term 'state' or
'stative' in a broader and less language-dependent way: sitting and
standing are generally classified as states, even though the corresponding English verbs fail Vendler's definition. The latter thus
covers only limited-control states, i.e. those in which one can
remain even while asleep (Vendler repeatedly uses this test of
"statehood"). 'Limited control' means, first of all, that if all
the preconditions for a state are met, one cannot help being in it:
if you learn or are told the answer you cannot help knowing what it
is. Secondly, the state of knowing the answer lasts until you for-
get it, which is not (treated by language as) a freely chosen action. This points to another frequent feature of limited control:
once you're in a state, you can't do something that will directly
cause its termination. Section 5 indicates how the two features of
limited control are, contrary to Vendler's opinion, directly connected to temporal reasoning.

Predicates describing limited control states with stable preconditions possess a property that has received a good deal of attention:
if such a predicate is true over a time interval it is true over
every subinterval of that interval. (Dowty 1986) uses this property
to define statives (giving, surprisingly, sleep as an example, although we can truthfully say that Billy slept from 10 to 6 even if
he got up a couple of times to go to the bathroom). I believe this
property to be a consequence of limited control and stable preconditions: if the preconditions hold until something happens and automatically insure that the state holds, then the state holds continuously. Also, there is no time bound on how long the state lasts:
if the preconditions hold while the person is eating or asleep then
the state can last forever. It is in this crucial respect that
limited control states differ from such states as standing, sleeping, or being drunk. Preconditions for these states include resources which get gradually exhausted: standing requires patience
and physical strength, sleeping requires a certain amount of sleepi-
ness or tiredness which gradually wears out, causing the sleeper to
wake up, and drunkenness wears out also unless replenished. In the
absence of disturbances from outside the system, such as an alarm
clock, the duration of the state monotonically increases with the
amount of resource, as in She was so tired that she slept until noon
or She slept meagerly during the week but caught up on weekends.

4.2. Processes. The notion of a consumed resource is even more
relevant for processes than states. Resource-consuming states all
seem to require only generic internal resources, which are not
specific to any given state but rather to all individuals of a given
sort. Within processes, there are those that require only generic
resources (walking) and those that require process-specific
resources as well: reading, for example, requires not only being
awake and not too hungry, but also a text to read. Telic processes
can be defined as processes that consume a specific amount of a
domain-specific resource. Resources are understood broadly: walking from A to B consumes the distance between them, building a house

consumes the as-yet-unbuilt but envisioned part of it, and destroying a house consumes the finite amount of 'structure' or 'order'
built into it. These examples illustrate three main classes of
telic processes: creating an object, destroying an object, and moving a specified amount of material (possibly the mover himself) to a
specified destination. Moving is understood to include all three
of Schank's PTRANS, ATRANS and MTRANS classes, with the proviso
that, unlike physical motion, MTRANS really copies structures from
the source to the destination. Moving also includes gradual (but
not instantaneous) changes of state.

4.3. Instantaneous events. 'Instantaneous events' is, of course, a
misnomer for those h-types whose h-tokens are all instantaneous,
i.e. have lifetimes that are very short closed intervals. The
linguistic test for verbs that describe instantaneous events is that
the imperfective or progressive of such verbs is either non-existent
or cannot be used to put the Window inside the history. The meaning
of the imperfective or progressive forms, if they do exist, turns
out to be different for different classes of the taxonomy presented
in the table below, even though the classification is based on a
seemingly unrelated principle: instantaneous events are grouped
together according to how the world before them compares to the
world after them. Since an event can meet and be met only by an
open interval there are the following possibilities (E stands for
instantaneous event):

TABLE 1.   Instantaneous events

Configuration	Label	Examples
state-E-same state	happening	flash, cough, jump
state-E-different state	transition	forget, recognize, notice,
process-E-state	culmination	win the race, reach the top
process-E-same process	disturbance	
state-E-process	flip-on	
process-E-different process	switch	

Interestingly, English verbs describing instantaneous events all
seem to fall into the first three groups, where the instantaneous
event meets a state, while beginnings of processes are rarely lexi-
calized.

4.4. The meaning of the progressive and the "imperfective paradox."
There have been several treatments, of which Dowty 1979 is the lat-
est and most detailed, that attempted to define progressive as a
sentential operator that has a uniform meaning for all sentences.
Vlach 1981 is, I believe, the first to propose that the meaning of
the progressive depends on the lexical class in a taxonomy of the
sort developed here. This subsection improves on Vlach's treatment
in two areas: instantaneous events, where my taxonomy is more detailed, and telic histories, where a new approach to their
well-known logical problems is proposed.

Among instantaneous events, happenings always have a progressive
form, and they always describe a sequence of instantaneous h-tokens.
Transitions may or may not have a progressive form; if they do, progressive is again associated with plurality, e.g. He's been noticing
that... (the implication being that the noticing has taken place on
many occasions). Finally, culminations (Vendler's 'achievements'),
usually have progressive forms that allow of two interpretations.
One is the familiar idea of plurality: The travelers, one by one,
were reaching the summit. The other interpretation arises from the
nature of culminations, which crown processes with sharply defined
end-points: the top of a climb, the finish of a race. The progressive of a culmination verb describes an h-token whose lifetime is an
open interval properly contained in the time of the process that
leads to the instantaneous event.

Outside instantaneous events, the semantics of the progressive is
quite straightforward for atelic processes and notoriously complicated for telic ones. The difficulty, as (Dowty 1977:46) states it,
is "to give an account of how [John was drawing a circle] entails
that John was engaged in bringing-a-circle-into-existence activity
but does not entail that he brought a circle into existence." This
formulation misses, I believe, the crux of the matter, which is the
default, non-monotonic nature of the inferences involved. The progressive sentence puts the Window in the middle of an open interval,
the time of John's activity; if another sentence puts the Window
after that interval, we infer, unless or until told otherwise, that
the activity came to its natural end. More precisely:

> The Telic-Imperfective Inference rule: If one sentence in a
> narrative tells us that a telic history h1 has begun or is in
> progress, and a later sentence moves the Window into or beyond a
> later history h2, and the time distance between h1 and h2 is of
> the same or a greater scale than the time scale of h1, then as-
> sume, unless or until told otherwise, that the end-point of h1 is
> reached.

The rule can be illustrated by the following narrative: In the
afternoon, when Johnny was doing his homework, Mommy went to the
market. In the evening they played together. Knowing how long it
takes, we infer that Johnny did his homework and Mommy her shopping.
However, the inferences are withdrawn if contradicted: the narrative could continue with until Mommy discovered that Johnny never
finished his homework. To see the importance of time scale, replace
"doing his homework" with "reading War and Peace."

This discussion of the progressive can be made more general and language independent by positing the following question: If the narrative puts us in the middle of an h-token, what are the possible ways
in which we can expect it to come to an end? Leaving zero resource
states aside, the answer depends on the internal dynamics of the
h-token and the presence or absence of volitional control. If the
h-token has an Agent participant, it can be interrupted by the
Agent's decision to do so. Apart from volition, the h-token comes to
an end when (a) the process-specific resource runs out (natural

end); (b) a generic resource runs out (exhaustion); (c) the preconditions no longer hold (interruption). Consider the h-token described by Sharon was reading a novel. It may come to an end when
(a) Sharon finished the novel (natural end); (b) Sharon got tired of
reading (exhaustion); (c) something happened that made reading impossible: her friend took the novel away from her, the lights went
out, her attention was distracted by loud music (interruption); and
finally (d) Sharon decided to put the novel down (volition). Possessing a natural end is, of course, a definitive property of telic
processes.

5. THE LEXICON AND NAIVE PHYSICS

The last paragraph is an example of what is called limit analysis in
the Qualitative Process Theory (QPT) of (Forbus 1985). (Since QPT
is only concerned with physical systems, it does not make the distinction between exhaustion and natural end, or between generic and
process specific resources. The distinction is clearly needed for
describing biological systems with goal-directed behavior.) In this
section I present an informal example of a partial lexical entry,
borrowing freely from the representational ideas of the QPT. In
particular, an h-type describing a process is characterized by a set
of preconditions, quantity conditions and functional dependencies.
Preconditions can be terminated by changes outside the process; when
this happens the process is interrupted. Quantity conditions are
terminated by the dynamics of the process itself; depending on which
condition is terminated the process gets exhausted or comes to its
natural end. Functional dependencies relate various parameters of
the participants of the h-type. There are just two of them, q+ and
q-; x1 q+ x2 reads "x1 depends on and increases with x2", and similarly for q-.

The first part of a lexical entry locates the word meaning in the
isa hierarchy of the sortal predicates of the RL, including the
aspectual subtree developed in this paper. A phenomenon that frequently emerges is regular polysemy of a natural language word between classes of the hierarchy; for example, most English verbs
describing a process can equally well describe a telic or an atelic
one. This is illustrated by the following sketches for the meanings
of the verb read:

read-1 (R, Ws, L): atelic process of the MTRANS type

Read-1 is the basic process by which reader R converts the visual
patterns (characters and groups of characters) of the writing system
Ws into words and sentences of language L. (Most languages have
only one writing system, but some--Japanese, Romanian/Moldavian-have
more.) This meaning of read appears in such contexts as She can read
or Reading requires a certain maturity of mind.

Preconditions:
  Knows (R,L)
  Knows (R, Ws)

Can-see (R) ;touch-reading using the Braille alphabet is not
covered
Can-attend(R) ;a general precondition for MTRANS

All other kinds of reading (reading for the first time, re-reading,
proofreading, skimming, thumbing through) use **read-1** and share its
preconditions.

Quantity conditions:
Tiredness (R) < Exhaustion-point (R)

In combination with the general axiom that tiredness of an Agent, in
any h-token that has such a role, increases with time and reaches
the Agent's exhaustion point every 20 hours or less, this quantity
condition ensures a cut-off point for an uninterrupted stretch of
reading.

read-2 (R, Pt, Mt) telic process of the MTRANS type.

Pt and Mt stand for Physical and Mental text, respectively. A
physical text is an ordered set of physical characters of a Language
L, which are instantiations of the patterns of a writing system Ws.
Read-2 means to traverse the physical text Pt in the order stipulated by the writing system Ws, applying **read-1** to build an (imperfect) copy of Mt in the mind of R. To read-2 a text means to be in
between its beginning and end, moving towards the end:

Quantity conditions:
Position (R,Pt) > Beginning (Pt)
Position (R,Pt) < End (Pt)

Functional dependencies:
Position (R, Pt) q+ time-of (this-token-of-read-2)
Tiredness (R) q+ Difficulty (Mt)

The first dependency simply says that the more you read, the more
you have read. In combination with the preceding quantity conditions, it sets up a "natural end" for **read-2**. The second dependency
says that mental texts are characterized, among other things, by
their difficulty: one can be too tired to read Kant, but in perfect
shape for an Agatha Christie.

Influences:
Has-read (R, Mt)

These two sketches are meant to illustrate the idea that there is a
profound similarity between natural language understanding and the
enterprise of 'naive' or 'qualitative' physics. The similarity
goes well beyond the physical vocabulary of processes and states in
aspectual classifications, or physical metaphors for human conditions. I believe that the same mental model, operating with the
concepts of resource and energy, underlies our understanding of such
situations as <u>The car stopped because there was no fuel in the tank</u>,
<u>John was too tired to read</u>, and <u>We don't have enough people to fin-</u>

ish the project. It is the unity of the model that gives rise to metaphors like John/The project has run out of steam. It would be interesting to see how the knowledge representation schemes of naive physics need to be modified to become part of lexical entries for words describing processes and states and how the naive physics modes of inference can be employed for reasoning about the temporal structure of discourse.

It has recently been recognized that in reconstructing temporal relations between histories in a narrative, we take into account their aspectual class (Dry 1983, Dowty 1986). In fact, both aspectual and durational knowledge, as well as the telic-imperfective inference rule and the notions of preconditions and expended resources, all participate in this reconstruction. The entire apparatus of this paper is a preliminary for a larger, discourse-bound investigation.

BIBLIOGRAPHY

Allen, James, 1984. Towards a general theory of action and time, AI, 23, 123-54.

Bach, Emmon. 1986. The algebra of events. Linguistics and Philosophy, 9, 5-16.

Barwise, J. and Perry, J., 1983. Situation Semantics. Cambridge, MA: MIT Press.

van Benthem, J.F.A.K., 1983. The Logic of Time. Dordrecht: Reidel.

Bunt, H., 1985a. The formal representation of (quasi) continuous concepts, in Hobbs and Moore.

Comrie, B., 1976. Aspect. Cambridge: Cambridge University Press.

Dahl, O., 1985. Tense and Aspect Systems. Oxford: Basil Blackwell.

Dahlgren, K. and J. McDowell, 1986. Using commonsense knowledge to disambiguate prepositional phrase modifiers. Proceedings of the AAAI Conference, Philadelphia, 589-93.

Dowty, David, 1977. Toward a semantic analysis of verb aspect and the English 'imperfective' progressive, Linguistics and Philosophy, 1, 45-77.

Dowty, David, 1979. Word Meaning and Montague Grammar. Dordrecht: Reidel.

Dowty, David, 1986. The effects of aspectual class on the temporal structure of discourse. Linguistics and Philosophy, 9, 37-61.

Dry, Helen, 1983. The movement of narrative time. Journal of Literary Semantics, 12, 19-53.

Forbus, K. D., 1985. Qualitative process theory, in Qualitative Reasoning about Physical Systems (Ed. D. Bobrow). Cambridge, MA: MIT Press.

Hayes, P., 1978. The naive physics manifesto, in Expert Systems in the Microelectronic Age (Ed. D. Michie). Edinburgh, Scotland: Edinburgh University Press.

Hobbs, J. R., 1985. Granularity. Proceedings of the IJCAI Conference, Los Angeles, 432-5.

Hobbs, J. R. et al., 1986. Commonsense metaphysics and lexical semantics. Proceedings of the ACL Conference, New York, 231-40.

Hopper, P., 1978. Aspect and foregrounding in discourse, in Discourse and Syntax (Ed. Talmy Givon). Syntax and Semantics, vol 12. New York: Academic Press.

Mourelatos, A. P. D., 1981. Events, processes, and states, in Tedeschi and Zaenen.

Tedeschi P.J. and A. Zaenen, eds., 1981. Tense and Aspect. Syntax and Semantics, vol. 14. New York: Academic Press.

Vendler, Z., 1967. Verbs and times, in Linguistics and Philosophy. Ithaca, NY: Cornell University Press.

Vlach, Frank, 1981. The semantics of the progressive, in Tedeschi and Zaenen.

SEMANTICS FOR REIFIED TEMPORAL LOGIC.

Han Reichgelt

Department of Artificial Intelligence, University of Edinburgh,
Edinburgh, Scotland.

ABSTRACT

There are many reasons for wanting theories of time in AI, and given the arguments in favour of using logic, it seems only natural to turn to temporal logics. In this paper, I will discuss one approach to temporal logics, namely reified temporal logics. Reified logics are logics in which what would normally be propositions appear as terms. They can be regarded as formalisations in a first order language of the semantics of a modal logic. Reified temporal logics thus combine the expressiveness of modal temporal logics with the efficiency of first order theorem provers. Unfortunately, the best-known reified temporal logics, namely Allen's and McDermott's, have not been given an adequate model theory. This paper is an attempt to rectify this situation. I first define a modal temporal logic, and a model theory for it. Later I formalise the model theory for this modal logic in a reified temporal logic and define a model theory for this system.

INTRODUCTION

There are many reasons for wanting theories about time in Artificial Intelligence. First, planning systems have to be able to reason about time, especially in applications in which one action may interact with the effect of another. Second, since natural languages often have complicated tense systems, an adequate natural language system will need to be able to reason about time as well. Third, in certain expert systems applications, the system has to be able to reason explicitly about time. For instance, monitoring systems such as VM (Fagan, 1980), need to have this ability. Given the need for systems which can reason about time, and given the advantages of using logic in Artificial Intelligence, as argued by for example Hayes (1977) and Moore (1984), it seems only natural to turn to temporal logics.

Philosophers interested in the formal treatment of time have developed a large number of modal temporal logics. (For an overview, see for example Rescher and Urquhart, 1971). Modal temporal logics have the advantage of greater and more natural expressibility than the naive first order treatment of time, in which every n-place predicate is changed into an $n+1$-place predicate, the extra argument referring to the time at which the formula is supposed to be true. For example, under the naive treatment, sentences (1) and (2) would be translated be as (3) and (4) respectively, where NOW is a special constant referring to the present time.

(1) In the past, some suffered from TB.
(2) Some suffered from TB in the past.
(3) (Et)[(t < NOW) & (Ex)[TB-sufferer(x,t)]]
(4) (Ex)(Et)[(t < NOW) & TB-sufferer(x,t)]

But, in first order logic (3) and (4) are equivalent. Intuitively speaking, however, the preferred interpretations of the two English sentences are not equivalent. The first sentence is preferably interpreted as:- in the past, there were some people (who may not now exist any more) who suffered from TB, whereas the second sentence is preferably interpreted as:- there are some people (now) who suffered TB in the past. The problem arises because the set of individuals may change from one time to the next. I will refer to this problem as the problem of changing ontologies.

A second advantage of modal temporal logic is the fact that sentences like (5). containing multiple modal operators, are more naturally expressed in modal temporal logics than in first order temporal logics.

(5) Some time ago, John still hadn't arrived.

A possible context in which this sentence could be used is the following:- speaker and hearer are expecting John. The hearer has asked the speaker whether the speaker knows if John has arrived. The speaker answers that some time ago, John still had not arrived. In other words, the speaker answers that some time ago John's arrival was still in the future. In a modal temporal logic, (5) would be translated as something like (6), where in a first order temporal logic, it would be translated into (7). It seems to me that (6) is a far more natural way of expressing the English sentence in question.

(6) (PAST(FUTURE(arrives(john))))

(7) $(Et)(Et')[(NOW < t) \& (t' < t) \& arrives(john,t')]$

The main problem with modal logics for an AI point of view is that, unlike for first order logics, nobody has yet succeeded in writing efficient theorem provers for arbitrary modal logics. Most of the discussions in the literature are either restricted to the propositional case (e.g. Halpern and Moses, 1985) or make specific assumptions about the set of individuals over which one quantifies, which make it impossible to express the distinction between (1) and (2) above. (see Abadi & Manna, 1986, and Konolige, 1986, for proposals of this kind and Reichgelt, 1986, for a more detailed critique).

We thus face an impasse:- on the one hand, modal temporal logics are more expressive and more natural, whereas first order temporal logics, on the other hand, have a computational edge. It is possible to complicate the naive first order treatment to get around some of the problems associated with it (Cf Reichgelt, 1986), but in this paper I want to concentrate on another possible solution to this impasse, namely by generalizing Moore's (1985) work on epistemic logics to temporal logics. The reasons for preferring this approach over naive first order theories will become clear in section 3.5.

Moore's proposal was to reason in the meta-language[1] of some epistemic logic, which was Moore's main interest. So, rather than reason with formulas in an epistemic logic, Moore proposed to reason about those formulas by reasoning with meta-level formulas that formalised the interpretations of the object-level formulas in a possible world semantics for epistemic logic. Since the model-theory can be formalised in a first order language, one can combine the expressive advantages of modal treatments with the computational advantages of first order treatments. Following Shoham (1986), we will call systems of this kind "reified logics", reified because one considers as things and quantifies over what in normal modal treatments would be propositions.

---

[1] I use the term "meta-language" in the logician's sense as a language in which one can talk about the (logical) object language in which one is primarily interested. The term "meta-knowledge" and related terms such as "meta-level" and "meta-language" have also become generally used in the context of expert systems, where meta-knowledge is defined as knowledge that the system has about its domain knowledge.

It is interesting to see that the formal temporal systems developed by Allen (1981, 1982) and McDermott (1982) can also be interpreted as reified temporal logics. Indeed, McDermott explicitly does so. Although I will argue in section 3.5. that this interpretation is probably not entirely adequate for Allen's system, let us accept it for the moment. Shoham (1986) rightly points out that no adequate model theory has been defined for either of these formal systems, and that as a consequence the formalisms remain without a clear interpretation. Shoham tries to put this right in his paper, and he defines a model theory for a temporal logic that is very much in the Allen and McDermott style. However, his definitions suffer from some shortcomings to which I will return section 4. In this paper, I will give an alternative model-theory that avoids these problems.

The outline of the paper then is as follows:- because a reified temporal logic is a first order formalisation of a modal temporal logic, I will first give a precise definition of a modal temporal language TM, and a model theory for it. I will then formalise the model-theory TM in a language of a reified temporal logic TR, for which I will then define a model theory. Finally, I will compare TR with Shoham's system.

MODAL TEMPORAL LOGIC

In this section I introduce a modal temporal logic. Its purpose is to show what a model theory for a modal temporal logic looks like. The model theory will be formalised in a first order language in section 3. I am not going into the mathematical properties of the logic in question. Readers interested in these matters are referred to Rescher and Urquhart (1971). I should stress that nothing in what follows is new from a logician's point of view.

A modal temporal language, TM. In this section, I define the modal temporal language TM. In certain respects, TM is slightly out of the ordinary. Unlike the more standard modal temporal languages, in TM we can refer to specific points in time. Most standard modal temporal languages only have P (meaning "sometime in the Past"), F ("sometime in the Future"), H ("Has always been the case") and G ("is always Going to be the case"), and do not allow reference to specific points in time. This makes them expressively rather poor. The following defines the language TM.

The vocabulary of TM is the union of
1. the set of constants $C = \{a_1, a_2,..\}$
2. the set of variables $V = \{v_1, v_2,..\}$
3. for at least one n > 0, the set of n-ary predicate symbols $P^n = \{p_1^n, p_2^n, ... \}$
4. the set of time constants $TC = \{t_0, t_1,..\}$

The set of operators, OP, is defined as:-
1. F,P are operators
2. if $t \in TC$, then AT(t) is an operator.

The set of well-formed expressions of TM, WFF^{TM}, is defined as:-
1. if $i_1,..,i_n \in C \cup V$ and $p \in P^n$ then $p(i_1,..,i_n) \in WFF^{TM}$
2. if $p,q \in WFF^{TM}$, then $(p \rightarrow q)$, $(\neg p) \in WFF^{TM}$
3. if $p \in WFF^{TM}$ and $x \in V$ then $(\forall x)[p] \in WFF^{TM}$
4. if $p \in WFF^{TM}$, and $O \in OP$, then $(Op) \in WFF^{TM}$

The connectives v, & and ↔ and the existential quantifier $(\exists x)$ can be defined in terms of →, ¬ and (x) in the normal way. Also the operators H and G can be defined as follows:-

Hp := ¬P¬p
Gp := ¬F¬p

In what follows, I will omit brackets wherever this can be done without the risk of ambiguity. In particular, outermost brackets will be left out.

TM-structures. In this section we will define the notion of an interpretation for TM.

A TM-structure is a quintuple [I,T,<,n,ind] where
I and T are disjoint non-empty sets,
< is a partial order relation on T,
n ∈ T,
ind is a function from T into the power set of I, i.e. a function which assigns to
   every t ∈ T a I' ⊆ I, such that for every i ∈ I, there is a d ∈ Range(ind)
   with i ∈ d.

In the above definition, I and T are the sets of possible individuals and temporal states respectively. The requirement that both sets be non-empty and that the two sets be disjoint are obvious.

Note that we have not put any constraints on the set T. In particular, we have not specified whether T is a set of points, or a set of intervals. There are various ways of viewing time, and one of the basic choices is whether one regards points in time or stretches of time as basic (for a discussion, see part I of Van Benthem, 1982). In this paper, I will not go into this question. A TM-structure is compatible with either choice.

The intuitive meaning of the partial order relation < on T is the earlier-than relationship. < is therefore (at least) a partial ordering. Again, there are no further constraints on <.

Intuitively, n is the present time. In the definition of a TM-interpretation, n plays the special role that all formulas are interpreted with respect to it. Again note that there are no additional requirements on n. In order to make the above definition of a TM-structure more in accordance with our intuitions about time, one probably has to add some conditions on the behaviour of n with respect to <. In particular, even if one does not want to insist that there is exactly one future, but rather that there are many different possible future courses of events, there is still only one past (Cf McDermott, 1982). Thus, the restriction of the <-relation to the points in time before n has to be a total ordering. There are probably many other requirements one could think of, but I will not go into this question here.

The final element in a TM-structure is the function ind. It is a function from temporal states to sets of individuals. Intuitively, it assigns to every point in time, the set of individuals existing at that point in time. This function makes it possible for modal temporal logics to cope with the problem of changing ontologies, i.e. the problem that the set of individuals over which one quantifies need not be the same from one moment in time to the next. Thus, while the set I is the set of possible individuals, the function ind assigns to every point in time the set of possible individuals actually existing at that point. The additional requirement on this function is there to ensure that every individual in a structure exists at at least one point in time. Thus, an individual is only allowed to be as a possible individual if it exists at some point in time. Again, there are many additional requirements that might be put on the function ind, depending on ones view of individuals. For example, is the same individual allowed to exist at a given time, then to disappear out of existence at a later time, and to come into existence again at some time after that? If this is not allowed, then some extra requirements on ind have to be added.

Interpretations for TM. In this section, I will use TM-structures to provide interpretations for the language TM. The following defines the notion of an interpretation for TM.

An interpretation for TM or a TM-interpretation, is a pair [S,f] where S = [I,T,<,n,ind] is a TM-structure and f is a function from the set of basic expressions of TM, except the variables, to objects constructed by set-theoretical means out of I and T such that

if c ∈ C, then f(c) is a function from T into I such that for every t ∈ T, f(c)(t) ∈ ind(t).

if p ∈ P^n then f(p) is a function from T into the power set of I^n such that for every t ∈ T, f(p)(t) ∈ ind(t)^n.

if t ∈ TC, then f(t) ∈ T.

Again, there are a few remarks to be made about this definition. The first clause assigns to constants functions from points in time to individuals. The extra clause ensures that the individual to which a constant a refers at a given time at that point in time. Those familiar with modal logic will see that unlike Kripke (1972), I do not treat constants as rigid designators.

The second clause assigns interpretations to predicates. Again, the extra clause is there to ensure that the individuals which stand in some relationship to each other or have some property at a point in time, actually exist at that point. The third clause is straightforward.

Now that I have defined the notion of a TM-interpretation, we can turn our attention to the central definition, namely that of a TM-sentence being true in a TM-interpretation. I first define the notion of a sentence being true in a TM-interpretation under a value assignment g. A value assignment is a function from the set of variables of TM into the set D of domain individuals[2].

Let p be a TM-sentence, let M = [[I,T,<,n],f] be a TM-interpretation, let g be a value assignment. Then p is true in M under g, symbolically M $\models_g$ p, is defined as follows:-

if p is of the form q(k_1,..,k_n), then M $\models_g$ p, if <h(k_1),..,h(k_n)> ∈ f(q)(n), where h(k_i) = f(k_i) if k_i ∈ C, and h(k_i) = g(k_i) if k_i ∈ V.

if p is of the form (p1 → p2), then M $\models_g$ p, if M $\not\models_g$ p1, or M $\models$ p2.

if p is of the form (¬ p1), then M $\models_g$ p, if M $\not\models_g$ p1.

if p is of the form (∀x)[q], then M $\models_g$ p, if for all value assignments h, such that h(y) = g(y) for all variables y except possibly x[3], M $\models_h$ q

if p is of the form Pq, then M $\models_g$ p, if there is some M′ exactly like M except that n′ < n, and M′ $\models_g$ q

---

[2] Those familiar with Kripke semantics for modal logic will notice that I define the notion of truth in a model directly. In general, one first defines the notion of truth in a world, specialises that notion to the notion of truth in a model and finally defines validity.

[3] The qualification "except possibly x" might need some clarification. We are interested in all the possible values that can be assigned to x. Therefore, the only value assignments that are of importance here are those that assign the same value as g to all variables other than x. So, they may only differ in the value assigned to x. But we are interested in all the possible values of x including the one that g assigns to x. Therefore, the value assignments that we are interested in do not have to assign another value to x. Hence, the qualification.

if p is of the form Fq, then M $\models_g$ p, if there is some M´ exactly like M except that n < n´, and M´ $\models_g$ q

if p is of the form AT(t)q, then M $\models_g$ p, if there is some M´, exactly like M except that n´ = f(t), and M´ $\models_g$ q

The above definition is relative to a value assignment, i.e. given one way of assigning individuals to variables, it is possible to determine what the truth value of a formula is. The following definition is independent of any value assignment, and therefore defines truth of a proposition as such. The reader will notice that given these definitions, free variables are interpreted universally.

A sentence p is true in a TM-interpretation M, M $\models$ p, if for all value assignments g, M $\models_g$ p.

REIFIED TEMPORAL LOGIC

In the previous section, I discussed a modal temporal logic. In it, I defined the syntax of the language TM and a model theory for it. The language in which the semantics for TM was formulated, TM's meta-language, was a first order language. Reified temporal logics are based on this observation. The basic idea is to formalise TM's meta-language using a sorted first order logic, and to reason in this new logic. The advantages are that one has a first order theory, while getting the full expressive power of a modal language.

The outline of this section is as follows. First, I define the language TR of a Temporal Reified logic. Then I define a model theory for TR. The discussion may be obscure in places. However, whenever things get too complicated, readers are advised to remember that TR is basically a formalisation of the model theory for TM.

The language TR. Since TR is a meta-language for the modal language TM, we have to be able to speak about two types of entity. On the one hand, we need to be able to talk about expressions of TM. On the other, we need to be able to talk about the entities to which the expressions of TM refer. Thus, TR is most conveniently defined as a typed first order language. The type structure of TR is as follows. There are two main types, namely exp and den, standing respectively for expressions and denotations, each of which has some sub-types. The sub-types of exp are p and i, proposition expressions and individual expression respectively. There are three types of individual expression, namely t, c and v, time-constants, individual constants and variables[4]. The sub-types of den are t-den and d-den, namely t-den for points in time and d-den for domain individuals. The type hierarchy can be represented as follows:-

---

[4] Note that variables of TM are treated as TR-constants. This reflects the fact that at the meta-level, object-level variables are regarded as just another type of expression, more or less on a par with object-level constants. I realize that there is at least a potential for confusion here, but this is the correct way of doing things.

universal
    expressions (exp)
        proposition expressions (p)
        individual expressions (i)
        time expressions (t)
        constant expressions (c)
        variable expressions (v)
    denotations (den)
        points in time (t-den)
        domain individuals (d-den)

For every one of the above types, TR will contain both constants and variables[5].

We can partially construct the vocabulary of TR out of the vocabulary of TM. For each time constant in TM, we have a constant of type t in TR. Each individual constant in TM corresponds to a constant of type c in TR, and each variable in TM corresponds to a constant of type v. Finally, each n-ary predicate symbol in TM becomes an n-ary function symbol in TR with arguments of type i, and value of type p. The last clause reflects the fact that predicates in TM take individual expressions to make sentences. Thus, there is a TR-function symbol corresponding to each TM-predicate taking TR-expressions denoting TM-terms, (i.e. individual expressions of type c or type v) to give TR-expressions denoting TM-propositions.

In addition to the above vocabulary which can be constructed from the language TM, we have the special constant NOW of type t in TR. TR also contains three two-place function symbols with as arguments and values expressions of type p, namely AND, IF and OR. There is also three one-place function symbol NOT, PAST and FUTURE, whose argument and value are both of type p as well. We have two two-place function symbols FORALL and THEREIS, whose first argument is of type v, whose second argument is of type p, and whose value is of type p as well. Finally, we have a two-place function symbol AT, taking as arguments a expression of type t and an expression of type p, to give an expression of type p as result. Clearly, these special function symbols are the counterparts in TR of the logical connectives, quantifiers and operators of TM.

Given the above set of constants and function symbols, the set of terms of TR, TERM^{TR}, is defined in the usual way.

Because TR is a first order language we have the logical connectives &, v, ¬ and → and the quantifiers (∀x) and (∃x). Also, TR has four predicate symbols HOLDS, <, T-DENOTES and DENOTES. HOLDS is a two-place predicate. Its first argument is of type p, while it second argument is of type t. Its intuitive reading is that the first argument is true at the time denoted by the second argument. Thus, HOLDS is the counterpart in TR of the double turn-stile (⊨) in the interpretation of the modal language TM. < is also a two-place predicate, both of whose arguments are of type t. < is the symbol for the ordering relation on times. T-DENOTES is a two-place predicate whose first argument is a time-expression and whose second argument is a point in time. T-DENOTES is the counterpart of that part of an interpretation function in a TM interpretation which assigns a point in time to time constants. DENOTES is a

---

[5] In the previous footnote I already mentioned a possible source of confusion because I treated TM-variables as TR-constants. There is another potential source of confusion here. In addition to TR-constants corresponding to TM-variables, there are also TR-variables corresponding to TM-variables. The confusion should disappear when one realizes that TR-variables are in fact what logicians call meta-variables, variables in the meta-language which stand for an arbitrary expression in the object-language.

three-place relation whose first argument is of type c, whose second argument is of type d and whose third argument is of type t. DENOTES is the TR counterpart of that part of the interpretation function in a TM interpretation that assigns domain individuals to constants.

Given the above set of predicates, connectives and quantifiers, and given the definition of TERM^{TR}, the set of well-formed formulas of TR, WFF^{TR}, is defined in the usual way.

TR-structures. TR is a complicated first order language. This complication is reflected in the notion of a TR-structure. Because TR is a formalisation of the meta-language of TM, there have to be counterparts of TM-expressions in a TM-structure. On the other hand, in a meta-language one also talks about the denotations of the object-language expressions. Therefore, a TR-structure should also contain TM-denotations. Given these preliminaries, the following definition of a TR-structure, although complicated, should be comprehensible.

A TR-structure is a $<C_d, V_d, TC_d, P_d, I_d, T_d>$ where $C_d, V_d, TC_d, P_d, I_d, T_d$, are non-empty and mutually disjoint sets. T_d contains the special element t_0.

The set C_d is the set of denotations of constants of type c. It is the counterpart in a TR-structure of the constants in TM. Similarly, the sets V_d, TC_d and P_d are the counterparts of the variables of TM, the set of time-constants and the set of TM-sentences. The sets I_d and T_d are the sets of real individuals and points in time respectively. The distinguished element t_0 of T_d is the interpretation of the special time-constant NOW and is the TR-counterpart of the point in time n in the definition of a TM-interpretation.

TR-interpretations. In this section, I will first define the notion of a TM-interpretation. Then, I will define the notion of truth in a TM-interpretation. The actual definition in this section does not reflect all the intuitions. For a fully satisfactory definition there would have to be some complications. I will ignore these in this section but I will return to them in the next section.

A TR-interpretation is an ordered pair [M,f] where $M = <C_d, V_d, P_d, I_d, T_d>$ is a TR-structure and f is an interpretation function such that

if c is a constant of type τ, then f(c) is a member of the corresponding set of M, where the following are corresponding sets, with type p we have associated P_d, with type c we have associated C_d, with type v we have associated V_d, with type c we have associated TC_d with type t-den we have associated T_d, and with type d-den we have associated I_d.

if m is an n-place function symbol with associated type information then f(m) is a function from the cartesian product of the sets corresponding to the types of the arguments of m into the set corresponding to the type of the value. f(NOW) = t_0

if P is an n-place predicate whose arguments are of types $\tau_1, .., \tau_n$, then f(P) is an element of the n-ary cartesian product of the sets corresponding to each τ_i.

In the remainder of this section I need the notion of a TR-value-assignment. I will first define the notion of truth in a TR-interpretation with respect to a TR-value-assignment, before defining the more general notion.

A TR-value-assignment g is a function which assigns to each variable of type τ an element of the corresponding set in a TR-interpretation.

Given the notion of a TR-value-assignment, we can now define the notion of truth in a TR-interpretation with respect to a TR-value-assignment in a relatively straightforward manner.

Let S be a TR-interpretation, $<M,f>$, let g be a TR-value-assignment, then a TR-proposition p is true in M under g, $S \models_g p$, if the following hold:-

if p is of the form $pred(arg_1,..,arg_n)$, then $S \models_g p$ if $<h(arg_1),..,h(arg_n)> \in f(pred)$ where $h(arg_i) = g(arg_i)$ if arg_i is a variable and $arg_i = f(arg_i)$ otherwise.

In addition to these clauses, we have the normal clauses for conjunctions, disjunctions, negations, implications and universally and existentially quantified formulas.

Some complications. The above system, baroque as it may appear to be, is still not complicated enough to capture all the intuitions. In particular, there will have to be more requirements on the denotation of HOLDS. In this section, I will discuss some of the necessary complications.

The main need for additional complications arises from the interaction between the HOLDS-predicate and the special function symbols, AND, OR, IF, and NOT. If AND, OR, IF and NOT are to be the TR-counterparts of the conjunction, disjunction, implication and negation in TM, then the following sentences have to be valid, whereas they are not in a TR-interpretation as defined above. The notation (x:anything)[P(x)] is used for "for all x of type anything, P(x)".

$(\forall t:t\text{-den})(\forall p:p)(\forall q:p)[HOLDS(AND(p,q),t) \leftrightarrow (HOLDS(p,t) \& HOLDS(q,t))]$

$(\forall t:t\text{-den})(\forall p:p)(\forall q:p)[HOLDS(OR(p,q),t) \leftrightarrow (HOLDS(p,t) \vee HOLDS(q,t))]$

$(\forall t:t\text{-den})(\forall p:p)(\forall q:p)[HOLDS(IF(p,q),t) \leftrightarrow (HOLDS(p,t) \rightarrow HOLDS(q,t))]$

$(\forall t:t\text{-den})(\forall p:p)[HOLDS(NOT(p),q) \leftrightarrow \neg HOLDS(p,t)]$

The obvious way to ensure that the above propositions are indeed valid would be to put additional requirements on the denotation assigned to HOLDS by the interpretation function. For example, in order to make the first of these formulas valid, the following additional requirement should be put on f(HOLDS):- if $<f(AND(p,q)),t> \in f(HOLDS)$, then $<f(p),t> \in f(HOLDS)$ and $<f(q),t> \in f(HOLDS)$. Similar rules can be added corresponding to the other formulas.

A second set of additional requirements on the interpretation of HOLDS arise because of the interaction of HOLDS and the special function symbols, THEREIS and FORALL. In order to formulate these criteria, we need to introduce another three-place function symbol SUBST, whose first argument is of type v, whose second argument is of type c, and whose third argument is of type p. Its value is of type p as well. The intuitive interpretation of SUBST(v,c,p) is the proposition obtained from the proposition denoted by p be replacing all occurrences of the variable denoted by v by occurrences of the constant denoted by c. Given this function symbol, we can now formulate two more TR-sentences which we would like to come out as valid.

$(\forall p:p)(\forall t:t\text{-den})(\forall v:v)[HOLDS(ALL(v,p),t) \rightarrow$
$(\forall i:i\text{-den})(\exists c:c)[DENOTES(c,i,t) \rightarrow HOLDS(SUBST(c,v,p),t)]]$

$(\forall p:p)(\forall t:t\text{-den})(\forall v:v)[HOLDS(THEREIS(v,p),t) \rightarrow$
$(\exists i:i\text{-den})(\exists c:c)[DENOTES(c,i,t) \& HOLDS(SUBST(c,v,p),t)]]$

Again, it is relatively straightforward to formulate the additional requirements on the denotation of HOLDS. We leave this as an exercise for the reader.

The reason for taking this complicated treatment of the function symbol THEREIS, rather than defining HOLDS(THEREIS(x,p(x)),t) as something like (Ex)(HOLDS(p(x),t), has to do with the problem of changing ontologies. The sub-formula DENOTES(c,i,t) ensures that quantification is not always over the same set of individuals. We insist that the quantification is only over the set of domain individuals that are denoted by some constant. If we identify the set of denoted domain individuals at some time t with the set of existing individuals, then we see that because we quantify over the set of denoted individuals, and because this set may be different between different points in time, we quantify over different sets of individuals depending on the time of evaluation. If we had taken the simpler option and defined HOLDS(THEREIS(x,p(x)),t) as something like (Ex)(HOLDS(p(x),t), then we would have been committed to accepting that the set of individuals over which we quantified, would always be the same. The quantifier would have been completely time independent. Therefore, we would not have been able to cope with the problem of changing ontologies.

A final set of extra requirements follows from the interaction between the function symbols PAST, FUTURE and AT, and the predicate HOLDS. In particular, we want the following three formulas to hold:-

$(\forall t{:}t\text{-den})(\forall p{:}p)[\text{HOLDS}(\text{PAST}(p),t) \rightarrow (\exists t'{:}t\text{-den})[(t' < t) \ \& \ \text{HOLDS}(p,t')]]$

$(\forall t{:}t\text{-den})(\forall p{:}p)[\text{HOLDS}(\text{FUTURE}(p),t) \rightarrow (\exists t'{:}t\text{-den})[(t < t') \ \& \ \text{HOLDS}(p,t')]]$

$(\forall t{:}t\text{-den})(\forall t'{:}t\text{-den})\forall(p{:}p)(\forall t\text{-exp}{:}t)$
$[(\text{HOLDS}(\text{AT}(t{:}\text{exp},p),t)] \ \& \ \text{T-DENOTES}(t\text{-exp},t')) \rightarrow \text{HOLDS}(p,t')]$

The intuitions behind the above formulas are again straightforward. We want the function symbols PAST, FUTURE and AT to behave as their counterparts in TM. It is again left as an exercise for the reader to formulate the additional requirements on the interpretation of the predicate HOLDS.

An important advantage of TR. TR has an important expressive advantage over TM and the naive first order treatment of time mentioned in section 1. The advantage, which is also pointed out by Shoham (1986)[6], is a direct consequence of the fact that one can quantify over propositions. In TM and in the naive theory, it is impossible to express general temporal knowledge such as "effects cannot precede their causes". Because one cannot talk about effects and causes in general, one can at most make the above statements about specific causes and effects.

In TR, one can in principle talk about causes and effects in general. TR contains constants and variables of type p, denoting expressions which in turn are naturally interpreted as denoting events etc. Because we can quantify over expressions of type p, TR gives one the power to talk about propositions and their denotations in general. Note that in the model theory of the present system there are no entities of type 'event'. That is, there is no TR-predicate P-DENOTES which relates propositions to appropriate denotations in a way which is similar to the way in which T-DENOTES relates time-expressions to points in time, and DENOTES relates individual expressions to domain individuals. Obviously, for the above proposals to work in an intuitive way, such a predicate would have to be introduced together with the required complication of the model theory. For example, one might to include events in the model theory as denotations for propositions. One can interpret Allen's work as a partial specification of the complications which are needed in the model theory for the above proposals to work. Under this interpretation, Allen's work should be regarded not as an exercise in reified temporal logic, but rather as an exercise in ontology. The question addressed by Allen

---

[6] Shoham does not seem to realize that the advantage here only holds for reified temporal logics. He seems to imply that it also applies to modal temporal logics.

is what complications are necessary in a model theory for a logical language that allows for the natural expression of temporal knowledge.

Note that the above proposal would also entail that TR can no longer be regarded as purely a formalisation of the meta-language of TM. After all, TM does not contain any expressions referring to events. But, even if we do not complicate matters any more, the present language is rich enough to allow us to express general temporal knowledge. For example, assuming that we defined a two-place predicate CAUSES, both of whose arguments are of type p[7], then we can express the above general temporal knowledge as:-

$(\forall p{:}p)(\forall q{:}q)[CAUSES(p,q) \rightarrow$
$(\forall t{:}t\text{-den})(\forall t'{:}t'\text{-den})[HOLDS(p,t) \ \& \ HOLDS(p,t') \rightarrow \neg \ (t' < t)]]$

As the above example shows, in order to be able to write down this piece of general temporal knowledge, it is imperative that we be able to quantify over propositions. Since this is not possible in TM, or in the naive first order treatment of time, this is a definite advantage of TR over either.

RELATED WORK.

The most relevant piece of related work is that of Shoham (1986) who also proposes a formalism that is also intended as a model theory for a reified temporal logic. Rather than discuss Shoham's formalism in detail, I will just point out the major differences between his system and the present system, and the main weaknesses of his system. For ease of reference, I will call Shoham's system SL.

First, it has to be admitted that SL looks far more elegant than TR, but, as I will argue, SL achieves this at the expense of less expressive power, and at the expense of a confusion between things that need to be distinguished. Sometimes, elegance has to be sacrificed for expressiveness and logical clarity.

The main superficial difference between SL and TR is that SL has a three-place predicate TRUE, whereas TR has a two-place predicate HOLDS. TRUE takes two time arguments and a third argument which is formed from putting what Shoham calls an n-place relation symbol in front of a sequence of n non-temporal terms, variables and constants. The intuitive reading is that the third argument is true between the points of time denoted by the first and the second argument. Shoham is not quite clear about the status of these third arguments but given the fact that he says that he is encoding possible-world semantics into the logic itself, and given his choice of the predicate TRUE, we have to assume that the third arguments correspond to expressions of type p. It is clear that this first superficial difference is minor and that changing the HOLDS-predicate into a three-place predicate would not pose insurmountable difficulties.

Shoham calls the expressions that are used to construct the third argument to the TRUE-predicate n-place relation symbols. This is an unorthodox use of this term. The expressions in question do not function as relation symbols normally do in logical languages. They are not used to construct atomic sentences. Rather, they are used to construct terms to the predicate TRUE, and are therefore more like function symbols. Although the above observation might seem pedantic, it is important because it points to a fundamental confusion in SL:- although SL is intended as a formalisation of a

---

[7] This shows that the predicate P-DENOTES should really be introduced. It is counter-intuitive to regard CAUSES as a predicate involving two propositions rather than as a predicate involving two events or whatever.

meta-language that has as its only (real) predicate the predicate TRUE, it does not distinguish between expressions denoting object-level expressions (i.e. in our terminology expressions of type exp) and expressions denoting real objects (in our terminology expressions of type den). Thus, in a sense SL confuses use and mention of object-level expressions. TR distinguishes between the two by having separate expressions for denoting object-level expressions and for denoting the real individuals which object-level terms denote. Also, TR has the predicate DENOTES for relating object-level terms to their denotations.

Bearing in mind this fundamental problem with SL, there are some other differences between SL and TR. SL allows only (the meta-level counterparts of) object-level atomic propositions as the third argument to TRUE. That is, SL does not contain the counterparts of the TR function-symbols AND, OR, NOT, IF, PAST, FUTURE, AT, THEREIS and FORALL. This is not disastrous as far as AND, OR, NOT, IF, PAST and FUTURE are concerned. After all, it was straightforward to define these function symbols using the formulas given in the previous section. Thus, we could define HOLDS(AND(p,q),t) as HOLDS(p,t) & HOLDS(q,t) etc. However, a similar treatment would not be possible for AT. In order to get an axiom defining the behaviour of this function symbol, the predicate T-DENOTES has to be introduced. Consider for example the proposition HOLDS(AT(t:exp,p),t). This proposition is true if the proposition denoted by p is true at the time denoted by t:exp. However, in order to formulate this, one has to be able to relate time-expression to points in time. However, this can only be done if the predicate T-DENOTES can be defined, and the possibility of defining T-DENOTES arises only when there is a distinction between time-constants and points in time. For related reasons, it is also impossible to give axioms defining the function symbols THEREIS and FORALL, at least not if we want to allow for changing ontologies. As already argued in the previous section if we defined HOLDS(THEREIS(x,p(x)),t) as (Ex)(HOLDS(p(x),t), then we would be committed to accepting that the set of individuals over which we quantified, would always be the same. The individual quantifier would be completely time independent. Thus, given this treatment of the function symbol THEREIS, HOLDS(THEREIS(x,PAST(p(x)),t) would be translated into (Ex)(Et')[(t' < t) & HOLDS(p(x),t')], while HOLDS(PAST(THEREIS(x,p(x))),t) would be translated into (Et')[(t' < t) & (Ex)[HOLDS(p(x),t')]]]. But these propositions are equivalent. Therefore, we would not be able to cope with the problem of changing ontologies.

Summarizing then, Shoham's system SL appears more elegant than TR but suffers from a fundamental confusion between use and mention of object-level expressions. Also, SL only allows atomic propositions as the third argument to the TRUE-predicate. Therefore, it cannot deal with the problem of changing sets of individuals.

CONCLUSION

Reified temporal logics (and reified modal logics in general) combine the naturalness of expression of modal temporal logics with the efficiency that can be achieved in theorem provers for first order logic. In addition, they have the important advantage of allowing the expression of general temporal knowledge of the type "effects cannot precede their causes", something which can be expressed directly in a modal temporal language. In this paper, I presented TR, a language for a reified temporal logic, and defined a semantics for it that avoided some of the problems associated with a similar system in Shoham (1986).

Acknowledgements:-
The research reported here took place as part of a joint Alvey project with GEC Research "A flexible toolkit for expert systems". I would like to thank Frank van Harmelen and Peter Jackson, as well as Barry Richards, for comments on earlier versions of this paper.

REFERENCES

Abadi, M. & Z. Manna (1986) Modal theorem proving. In J. Siekman (ed.) *8th International conference on automated deduction.* Berlin: Springer-Verlag.

Allen, J. (1981) A general model of action and time. TR 97, Dept of Computer Science, University of Rochester, Rochester, NY.

Allen, J. (1982) Maintaining knowledge about temporal intervals. *Communications of the ACM,* 26, 832-843.

Fagan, L. (1980) Representing time-dependent relations in a medical setting. Ph.D. Diss. Computer Science Department, Stanford University.

Halpern, J. & Y. Moses (1985) A quide to modal logics of knowledge and belief: Preliminary draft. IJCAI-85, 480-90.

Hayes, P. (1977) In defence of logic. IJCAI-77, 559-65.

Konolige, K. (1986) Resolution and quantified epistemic logic. In J. Siekman (ed.) *8th International conference on automated deduction.* Berlin: Springer-Verlag.

Kripke, S. (1972) Meaning and necessity. In D. Davidson and G. Harman (eds) *Semantics of natural language. (2nd edition)* Dordrecht: Reidel.

McDermott, D. (1982) A temporal logic for reasoning about processes and plans. *Cognitive Science,* 6,101-55.

Moore, R. (1984) The role of logic in Artificial Intelligence. SRI International. Technical Note 335.

Moore, R. (1985) A formal theory of knowledge and action. In J. Hobbs and R. Moore (eds) *Formal theories of the commonsense world.* Norwood, N.J: Ablex.

Rescher, N. & A. Urquhart (1971) *Temporal logic.* Berlin: Springer-Verlag.

Reichgelt, H. (1986) A comparison of first order and modal treatments of time. Technical report EdU-12.1 of "A flexible logic-based toolkit for expert systems".

Shoham, Y. (1986) Reified temporal logics:- Semantical and ontological considerations. ECAI-7, 390-97.

van Benthem, J. (1982) *The logic of time.* Dordrecht: Reidel.

TLP — A TEMPORAL PLANNER

Edward P K TSANG

Department of Computer Science
University of Essex
Colchester, U K
Electronic Address: edward@uk.ac.essex

ABSTRACT

In (Tsang 1986a), we presented a framework for planning. In this framework, knowledge of the world is modelled by reference to an interval-based time logic. Planning is seen as a knowledge manipulation mechanism, the goal of which is to generate plans/schedules which are ready for execution. This paper gives a brief report of TLP, an implementation of a planner in which such a mechanism is incorporated.

I INTRODUCTION

Conventional planners like STRIPS (Fikes 1971), NONLIN (as described in (Tate 1976,77,84)), and NOAH (Sacerdoti 1974,75) are limited in their ability to represent complex temporal relations, such as "activities A and B must finish at the same time" (see (Tsang 1986a)). A DEVISER-type (Vere 1983,85a,85b) point-based temporal planner can only consider one temporal relation at a time. For example, it cannot represent constraints like "P1 must be before or after P2" in one network, where P1 and P2 are time points. In (Tsang 1986a), we have presented a framework for planning which can reason with disjunctive temporal relations. In this paper, we report an implementation based on this framework. Limited by space, we shall not attempt to present the details of the design. The main objective of this paper is to give an overview of a temporal planner with which one can reason with disjunctive temporal relations, and generate complete plans.

I.1 Knowledge Representation

In this research, knowledge of the world is modelled by reference to a temporal logic. In this model, each assertion is associated with an interval. An interval is **explained** if it can be shown to hold. Between two intervals, there are 13 possible temporal relations, which are shown in Table 1 below.

TABLE 1 — Possible temporal relations between X and Y

Relation	Symbols	their inverse	temporal relation
			YYYYYYYY
before	<	>	XXXX
meet	m	m i	XXXXXXXXX
overlap	o	o i	XXXXXXXXXXXXX
start	s	s i	XXXX
equal	=	=	XXXXXXXX
during	d	d i	XXXX
finish	f	f i	XXXXXX

In this logic, each temporal relation between two intervals has to take one of these 13 primitive relations as its value. The value that a temporal relation can take is restricted by constraints like the *transitivity rules, propositional constraints, domain constraints,* etc. Transitivity rules are constraints on temporal relations which are inherited in the logic that we use. For example if A *meets* B and B *meets* C then A is *before* C. An example of a propositional constraint is that if P holds in T1 and not(P) holds in T2, then T1 and T2 must be non-overlapping intervals. Domain constraints are constraints which are applicable to certain domains, e.g. in the blocks world domain the interval in which clear(B) holds must not overlap with the interval in which on(A,B) holds. Intervals and their temporal relations form a complete simple network, which is called a **relation network**. (A network is *complete* if there exists at least one arc between any two nodes. It is *simple* if it is anti-reflexive and symmetric.) We shall not go into this logic here. Interested readers are referred to (Allen 1985) (Tsang 1986b,87).

Knowledge of the world is represented as **Partial World Descriptions** (PWDs), each of which is a package of partial information about a possible world. The term "possible world" here means a model of a world which *could* exist, but that does not mean that it must be a world which contains "the present". It could be a hypothetical world. For convenience, we shall call the possible world that we are currently looking at the world of the current context, and all partial descriptions of it **current PWDs**. A language called the *PWD Language* is defined for describing partial world models. We refer to our planning approach as a **PWD Approach**. In this approach, planning is seen as the mechanism for the manipulation of PWDs in order to generate a schedule. A schedule here means a set of intervals mapped onto a time line. An example of a schedule is given in figure 1.

Figure 1 -- Example of a schedule

In designing our knowledge representation schema, we first have to decide what to include in our PWDs. Then we can define our language for world descriptions. It is likely that the richer our language is, the more we can model the real world, but the more computation we need to process it. We have defined a **Partial World Description Language** (PWDL). In this language, a partial world is expressed by a property list. It allows us to talk about the intervals and their assertions, constraints on temporal relations, durations and resources, inference rules in the knowledge representation schema, preferences, etc. The formal specification of the language will not be presented here. Instead, we shall explain what is meant when the examples are given later.

I.2   The Planning Mechanism

Under our formalism, problem specifications, operators, constraints and plans are seen as PWDs. A problem specification is expressed as a PWD in which some intervals could be unexplained. Operator definitions are PWDs which are given names, and are therefore called **named-PWDs**. (The names are used as index to the definitions.) Applying an operator means merging its PWD into the world of the current context, and perfoming all the necessary inferences. Intervals are explained by this PWDs-merging and inferencing mechanism.

In (Tsang 1986a), we have argued that Allen's algorithm for planning is incomplete. We suggested a complete planning process (which attempts to consider disjunctive temporal relations) which consists of the following steps:

1. Accepting the problem

2. Explaining all the intervals in the possible world

3. Labelling all the temporal relations

4. Computing the start and end times of each interval.

At the end of this process, all the intervals are explained and mapped onto a metric date line. This will be a schedule which is ready for execution.

II   IMPLEMENTATION OF THE TEMPORAL AND REASONING MODULES

In this section we shall describe some application-independent modules of our system: the **Temporal Inference Engine** (TIE), the **Duration Handler** (DH), the **Resource Allocator** (REAL) and the **Partial World Description Processor** (PWDP). These modules will be described in the following sections.

II.1 The Temporal Inference Engine (TIE)

TIE is a domain-independent inference engine which is primarily based on Allen's algorithm (Allen 1983b). It accepts **temporal constraints** as input, and propagates their effects to other temporal relations. A temporal constraint on intervals X and Y is a constraint on the temporal relation between X and Y. It is denoted by R(x,y) in this paper.

One specific feature of TIE is that it can handle points as well as intervals. Points are treated as atomic intervals. We stipulate that if X and Y are both points, then they can only be equal, or one of them is before another. The "starting point" of an interval T is seen as the atomic interval (point) that *meets* T. The "ending point" of T is the atomic interval that *finishes* T. Such a stipulation has the nice characteristic that if interval T1 meets interval T2, then the ending point of T1 is the same as the starting point of T2. Besides, once the points ·are introduced into the system, they can be handled as normal intervals. Therefore, introduction of points will not make temporal reasoning more complicated. Table 2 shows the temporal constraints on R(x,y), depending what X and Y are.

TABLE 2   Default temporal constraints on R(x,y)

X	Y	default temporal constraint
interval	interval	[< m o fi di s = si d f oi mi >]
interval	point	[< fi di mi >]
point	interval	[< m d f >]
point	point	[< = >]

Another important feature of TIE is that the user can use reference intervals to limit the computation needed for constraint propagation. When the number of intervals in the problem grows, the number of temporal constraints in the database of TIE grows exponentially. Then the constraint propagation process would become very laborious. The basic idea of using reference intervals is to organize the intervals into a ·hierarchy, hoping that the relevant intervals are grouped together and the irrelevant ones are separated. An example of a hierarchy is shown in figure 2.

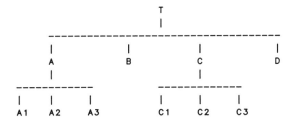

Figure 2   — Example of a hierarchy of intervals

A parent in a hierarchy is the *reference interval* of its immediate children.
We define two intervals A and B as **referable** to each other if they have the
same reference interval, or if one of them is the reference interval of the
other.   In the above example A and B are referable to each other; A1 and D
are nonreferable to each other.   Only the temporal constraints between
referable intervals are stored explicitly.

When a constraint between two referable intervals X and Y is input to the
system, it is only propagated among those intervals which are referable to
both X and Y.   When a constraint between two nonreferable intervals, for
example A1 and C1, is input to the system, Allen's planner would create a
link to make them referable to each other. In the above example, it may make
A an additional reference interval of C1, or C an additional reference interval
of A1 (or make one of A1 and C1 the reference interval of the other).
However, by so doing, the hierarchy will soon be 'flattened' (Allen 1983b),
which means soon most of the intervals will become referable to each other.
So the use of reference intervals cannot serve its purpose of reducing the
efforts needed in constraint propagation.

In TIE, when the constraint between A1 and C1 (i.e. R(a1,c1)) is imposed, it
will only be used to examine the temporal constraints R(a,a1), R(a,c) and
R(c,c1), (and update them when necessary).   Therefore, the effort needed for
constraint propagation can be reduced.   We must point out that when
reference intervals are used in this way, certain information can be lost in
TIE.   But this is tolerable, because the constraint propagation algorithm is
incomplete anyway.   Details of this argument will not be presented here, but
see (Tsang 1986a,87).

II.2   The Duration Handler (DH)

The Duration Handler (DH) is used as a front-end for duration constraints
reasoning.   A **duration constraint** is a constraint on the minimum or maximum
duration of an interval.   Like Allen's *Duration Reasoner*, DH cannot guarantee
consistency with regard to duration constraints (Tsang 1986a).   It uses simple
rules to detect inconsistency in relation networks, hoping that flaws which fall
into certain general patterns can be detected at an earlier stage.   The basic

rule that DH uses is:

> If T1, T2, ..., Tn are subintervals of T,
> AND T1, T2, ..., Tn are discrete from one another,
> THEN the summation of minimum durations of T1, T2, ... Tn
>   must be less than or equal to the maximum duration of T.

As Allen has pointed out, temporal and duration constraints are closely related to each other (Allen 1983b). For example, if the temporal relation between the intervals A and B is restricted to [=], then the constraints on the maximum and minimum durations of both A and B should be the same. As one might expect, TIE and DH communicate.

II.3   The Resource Allocator (REAL)

The Resource Allocator (REAL) is used to update the planner's knowledge of the future world concerning variable bindings. The user is allowed to declare hierarchies of resources (e.g. "a lathe is a kind of machine") and the resources available for a resource type (e.g. "there are two compressors, A and B"). REAL can also maintain arc-consistency of simple constraints on resources (e.g. "X must not be the same object as Y") (Mackworth 1977). But REAL is a very primitive implementation. It can only reason with certain resources -- namely resources which can be used and released (not, for example those which can be created, split or destroyed, see (Steel 1985)).

II.4   The Partial World Description Processor (PWDP)

The PWD Processor (PWDP) is used to interpret the PWD Language and add the information provided by the interpreted PWD into the current PWD. The slots in a PWD Language are processed as follows:

1. Assertions and intervals
    Under the slot *interval* is a list of intervals. PWDP will generate names for those uninstantiated intervals so that they can be referred to. Under the explained and unexplained interval slots are lists of pairs:
    [Assertion, Interval].
    When these slots are encountered, PWDP will do the following:

    a. generate a new name for each uninstantiated variable and interval so that they can be referred to in the future.

    b. Record that the Assertion holds in its corresponding interval. For those intervals which are declared under the slot *explained_interval*, record that they are "explained" by declaration.

    c. *Precondition* is treated as unexplained interval plus temporal constraints (constraints between the interval and the interval in which the operator holds). *Effects* and *side-effects* are treated as explained intervals plus temporal constraints. Temporal constraints are processed by TIE, which is described above.

2. reference_intervals & points
   This information will be recorded in the database. Reference intervals are used to build the hierarchy of intervals, and information about the points is used in temporal reasoning in TIE.

3. temporal_constraint, duration_constraint, resources
   These constraints are processed by TIE, DH and REAL respectively.

4. inference_from_proposition
   We allow inference rules to be defined in our PWD Language. Inference rules play an important part in our knowledge representation schema. It helps us to define conditional and repetitive operators. Discussion of it will be left for another occasion (see (Tsang 1987)).

5. preference_in_temp_rel, absolute_times
   Such information will simply be recorded. Preference will be used in definite plan generation. Absolute time will be used in the computation of the start/end times of the intervals.

6. named_WDs
   As mentioned before, operators are named-PWDs. A named-PWD can be identified by its name, arguments and intervals. For example:
   [cook, [john, fish], [T1,T2,T3,T4]]
   is the identifier for the operator definition of cook(john,fish), in which intervals T1, T2, T3 and T4 are involved. We stipulate that the first interval in the list refers to the interval in which the assertion holds. In this example, cook(john,fish) holds in T1. When this triple appears in the name_WDs slot of the PWD W, it means that the PWD declaration for the operator "john cooks fish" (with the variable and interval bindings) is part of W.

   Indexed by the name, arguments and intervals, PWDP searches for the corresponding PWD (operator) declaration. If it is not found, PWDP will report failure. If the operator is found, it will be input to PWDP recursively.

7. Remarks and slots under other names
   All slots which names are not mentioned above are treated as remarks. They are only meaningful to the user, and are ignored by the PWDP.

III   IMPLEMENTATION OF THE TEMPORAL LOGIC PLANNER

TLP is a domain-independent planner. The structure of TLP and its relation with the application-independent modules is shown in Figure 3 below. The modules of the planner are described in the sections that follow.

III.1   The Interval Explanation Module (IEM)

Input to this module is a Partial World Description (PWD) in which some intervals may be unexplained. The IEM attempts to explain all the intervals in the current PWD. The output of it is a PWD in which all the intervals are explained. The intervals and their temporal relations in this PWD output form a relation network.

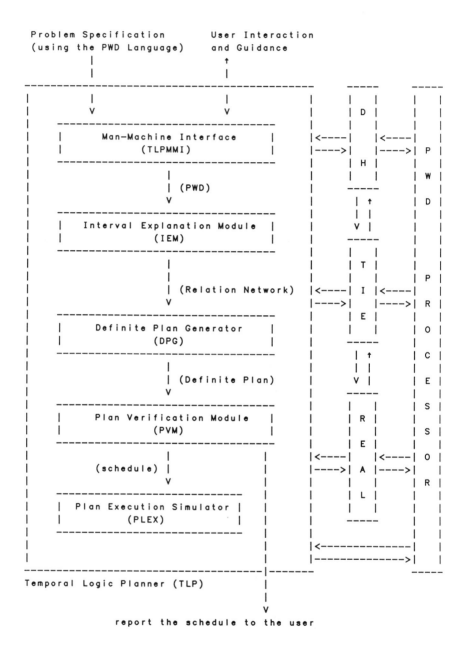

Figure 3   —   Architecture of TLP and its related modules

The following are possible means to explain an interval (for convenience, we call the interval that we want to explain T):

1. We can assert that interval T is the same interval as T', which is an existing explained interval in the current PWD. To be able to do so, the same propositions must hold in both T and T'.

2. If proposition P holds in T, then one possible way to explain T is to find an explained interval T' in which P' holds, such that P and P' unify. Then T and T' can be equated. For example, free(X) holds in T1, and free(john) holds in T2 where T2 is explained and T1 is not. In order to explain T1, we can (i) assert that X is john; and (ii) assert that T1 is the same interval as T2.

3. One way to introduce more explained intervals into the current PWD is by application of operators. As we have explained before, an operator is just a PWD. By merging the operator's PWD into the current one, we get more explained intervals, which can be used to explain T by applying (1) or (2) above.

When two intervals T1 and T2 are asserted to be the same, we are imposing temporal and duration constraints on our PWD. The temporal constraint imposed is T1 [=] T2. Consequently, both T1 and T2 must have the same duration constraints. These constraints will be handled by TIE and DH, which will report failure if the current PWD is over-constrained.

Choice points exist during this interval explanation process. Often, an interval T can be collapsed with more than one interval, or achieved by more than one operator. Besides, for example, free(X) may be unified with free(john), free(mary), or other predicates. IEM has to commit itself to one choice at a time, and backtrack (chronologically) to others if that choice is proved to be nonviable. Therefore, IEM is basically employing a look ahead search strategy -- using TIE, DH and REAL to detect inconsistency (Haralick 1980).

III.2   The Definite Plan Generator (DPG)

Input to the **Definite Plan Generator** (DPG) is the PWD output by IEM. DPG attempts to label the temporal relations in the relation network of this PWD. A **label** is a pair <R(x,y), r> -- the assignment of a primitive relation r to a temporal relation R(x,y). If the relation network is consistent, DPG will output a **definite plan**, a relation network which has each of its arcs labelled to a primitive temporal relation. If no consistent labelling is available, DPG will report failure.

Labelling the temporal relations is a **consistent labelling problem** (CLP). The nodes of the constraint network are the temporal relations, and the arcs are constraints -- such as the transitivity rules, propositional and domain constraints. The formal specification and analysis of the CLP in temporal reasoning is an interesting topic. Limited by space, it will be left for another occasion (see (Tsang 1987)). Some techniques developed for CLP are used in DPG, for example in the order of the labelling. When the relation network is built, TIE always maintains 3-consistency (Freuder 1978,82) in it. Maintaining a certain level of consistency prevents the planner from repeating wrong choices (Mackworth 1985). DPG picks one temporal relation at a time, and attempts one labelling at a time. TIE and DH are used to propagate the effect when each label is made. Therefore it is basically using a *forward checking* searching strategy, which, according to Haralick, is one of the most efficient searching strategies in a CLP (Haralick 1980b).

After all temporal relations in the relation network are labelled, all the intervals can be mapped onto a date line and reported as a schedule. However, this schedule shows only the relative temporal relations of the intervals, e.g. "A must be done during B". It does not tell, for example, at what time A and B should start, and how much earlier must A start before B.

III.3   The Plan Verification Module (PVM)

The **Plan Verification Module** (**PVM**) considers the metric properties of time in the relation network, which includes duration, waiting time, etc. It attempts to compute the start and end time of each interval. If absolute time is provided in the problem, it will be used in the computation; otherwise, relative time will be used. If the relation network is inconsistent with regard to metric time, the PVM will report failure.

In order to compute the start and end time of each interval, PVM has to build a graph of start and end points from the definite plan. The nodes of this graph are start or end points of the intervals in the definite plan. The arcs are directed, possibly associated with a time length, showing the precedence between the two points that it connects. If the minimum duration of interval T is N, then it builds an arc of length N from start(T) to end(T). If the maximum duration of T is X, then it builds an arc of length -X from end(T) to start(T). Since all intervals have been mapped into a date line, (hence they are linearized), each definite plan forms a unique graph of time points. After this graph has been built, PVM uses Bell and Tate's *AnalyzeLongestPath* algorithm to check whether there exists a feasible solution in our network. At the same time it will compute for each the node the earliest possible time if the network is consistent. The AnalyzeLongestPath algorithm is a specialized linear programming algorithm suitable for this application (Bell 1985b).

We said earlier that information could be lost in TIE if we use reference intervals in our way. The solution to that is, when the temporal constraint on two nonreferable intervals is processed, it is recorded by TIE. Such information is used by PVM to build the graph of time points. We argue that any inconsistency caused by the lost of information will be detected at this stage.

III.4   The Plan Execution Simulatior & Man-machine Interaction Module

A primitive **Plan Execution Simulator** (**PLEX**) has been built in TLP to demonstrate how plan execution can be fitted into the PWD approach. Besides, it provides a basis to test conditional and repetitive operators. During plan execution, new information can be input to the system. This will be handled by the same knowledge manipulation mechanism. That means new unexplained intervals will be detected, and new plans will be generated when necessary. The **Man-Machine Interaction** module (TLPMMI) is used by TLP to interact with the user, who can inspect the knowledge of the system as well as giving guidance. We shall not go into details of the PLEX and TLPMMI here.

IV EXAMPLES

The planner has been tested in various domains, such as cooking and job shop scheduling. In this section, we shall report one of the tests in the job shop scheduling domain.

A task in this domain is to construct a schedule for a workshop, given a set of jobs. Each job is a process of a particular type. Each process involves a list of steps which must be performed in a certain order. Constraints exist among the steps. For example, a machine can only be used for one batch of job at a time. Some jobs must be finished before others. The manager often wants to add additional tasks and constraints to existing schedules. A problem expressed in our PWD Language is shown in figure 4.

```
[explained_intervals: [[plan, window]],
 unexplained_intervals:[[process_1(batch1), T1],
                        [process_1(batch2), T2],
                        [process_2(batch3), T3]],
 temporal_constraints: [[T1, [s,=,d,f], window],
                        [T2, [s,=,d,f], window],
                        [T3, [s,=,d,f], window]],
 resources: [isin(compressor, [compressor_A])]
]
```

Figure 4 -- the problem specification of our example

In this problem specification, we state that there is only one explained interval which is called window. There are three intervals that we want to explain, each of which representing a job. Two jobs are of process type 1, and one of process type 2. The "temporal_constraints" states that all jobs must be done within the window. In this example, we declare that only one compressor is available, namely compressor_A. Such information is handled by the Resource Allocator (REAL). The definition of the operator *process_1* is shown in figure 5.

[name: process_1,
 argument: [Batch],
 interval: [T0, T1, T2, T3, T4, T5, T6],
 resources:[isa(X, compressor)],
 precondition: [[[o,fi,di,s,=,si,d,f,oi], charged(X), T1]],
 side_effect: [[[si], p(machine1,Batch), T2],
               [[di], p(machine2,Batch), T3],
               [[di], p(machine3,Batch), T4],
               [[di], p(machine4,Batch), T5],
               [[di], p(machine5,Batch), T6]],
 temporal_constraints: [[T2, [<,m], T3],
                        [T3, [<,m,mi,>], T4],
                        [T3, [<,m], T5],
                        [T4, [<,m], T5],
                        [T5, [<,m], T6],
                        [T5, [s,=,d,f], T1]]
 ]

Figure 5   —   Definition of the operator *process_1*

The operator process_1 takes one argument, Batch. Seven intervals (T0 to T6) are involved in it, where (by stipulation) T0 is the interval in which process_1(Batch) occurs. This operator definition states that there are 5 processing steps, which hold in intervals T2 to T6, needing machines 1 to 5. The list under "temporal_constraints" states that T2 must be followed by T3 or T4, which must both precede T5, which in turn precedes T6. T1 is an unexplained interval in which charged(X) holds, where X is declared as a compressor (under "resources"). It is further specified that the step in T6 must be performed while X is charged (the last temporal constraint). There is no "effect" list in this operator definition. If there is, it is used as the pattern by which this operator is invoked. In this example, the operator is invoked to explain *process_1(Batch)* for some *Batch*. Definition of the operator *process_2* is similar to process_1. It requires 4 process steps (instead of 5), which requires machine6, machine7, machine1 and machine8 in that order.

Given the problem specification in figure 5, IEM explains the intervals T1 to T3 in it. That means "applying" process_1 twice and process_2 once. When an operator is "applied", the PWD in its definition is merged into the world of the current context. In order to explain T2, process_1(batch1) is applied and the interval in which process(batch1) holds is asserted to be the same interval as T2. Then all the inferences are performed. Intervals T3 and T4 are explained by the same mechanism. After all the intervals are explained, DPG generates a definite plan, which is shown in figure 6.

```
window                        ----------------------------------------------
process_1(batch1)             ----------------------------------------------
p(machine1,batch1)            -----
start_up(compressor_A)  --------------
process_1(batch2)                   ------------------------------
p(machine2,batch1)            ----
p(machine1,batch2)            ----
p(machine3,batch1)                  ----
p(machine2,batch2)                  ----
process_2(batch3)                         ------------------------------
p(machine4,batch1)                        -----
charged(compressor_A)                     ------------------------------
p(machine3,batch2)                        -----
p(machine6,batch3)                        ----------------
p(machine4,batch2)                              -----
p(machine5,batch2)                                ------
p(machine5,batch1)                                      -------------
p(machine7,batch3)                                      ----
p(machine1,batch3)                                            ----
p(machine8,batch3)                                            ----
```

Figure 6   --   the initial schedule generated for our example

In this example, we assume that after the initial schedule is generated, the manager wants to add new tasks and constraints. The new requirement is shown in figure 7, where the names like interval456 under "temporal_constraints" refer to the intervals in the existing schedule.

```
[remarks :   'more job and constraints',
 unexplained_intervals : [[process_2(batch4), A4]],
 temporal_constraints :   [[interval456, [m], interval789],
                           [A4, [<,m,o,s,d], interval111]
 ]
```

Figure 7   --   Additional tasks and constraints

After this new requirement is input to the system, IEM will detect the unexplained interval (A4) and try to explain it. Then a new definite plan is generated. The final schedule has the same format as the one presented in figure 6 and therefore will not presented here. Then PVM is called to compute the start and end times of each step, which means giving the schedule a time scale.

The planner and the modules described in this paper have been implemented in PROLOG on DEC-10, under the ¯TOPS-10 operating system. The schedules produced in this domain all seem to be reasonable. After the additional constraints are added and processed, 29 intervals and 406 temporal relations are involved in this example. TLP takes 17 CPU minutes to generate the initial plan, and nearly 40 CPU minutes for replanning. But the time spent is not a good measure of the program's performance because the program (plus the Prolog interpreter) has access to less than 250K of core only. So hundreds of garbage collections have been performed in running the above example.

Reference intervals are not used in this example because we assume that the user wants to show all processing steps and their temporal relations in the schedule. Because of this, each temporal constraint has to be propagated throughout the whole database of temporal relations. Reference intervals help to improve the performance of the planner. In an example in cooking domain, 35 intervals are involved. But since reference intervals are used, it takes only 18 CPU minutes to generate our schedule.

V DISCUSSION AND SUMMARY

In this paper, we have reported a temporal planner which demonstrates the PWD approach to planning described in (Tsang 1986a). In this approach, we see planning as a knowledge manipulation mechanism, the goal of which is to produce schedules which are ready for execution. Under this approach, all problem specifications, operators, plans and constraints are seen as a partial description of a possible world. In problem specifications, there is nothing called "initial" or "goal" states. Goals are just intervals which we want to show exist in a possible future world.

The whole planning process consists of the following stages: accepting the problem, explaining all the intervals, labelling the temporal relations, and computing the start and end times of each interval.

We have built TIE, DH and REAL, which are modules for temporal, duration and resource reasoning. A PWD Language is defined for knowledge representation. The PWD Processor is built to interpret the PWD Language. We have also built TLP, a temporal planner which makes use of the above modules to generate schedules.

We claim that TLP is useful for application domains where complicated temporal relations have to be represented. It is built to reason with disjunctive temporal relations. The merit of the planner is in its modularity: each module tackles a well-defined problem. But the planner has certain limitations. The combinatorial explosion problem exists in different stages of the planning process. Choice points exist in IEM (in how to explain an interval) and DCG (in which label to choose first). Besides, its resource reasoning capability is very limited.

This research is directly related to Allen's work (Allen 1983a,83b,85). Related works are Vere's DEVISER (Vere 1983,85a,85b), Dean's TMM (the intelligent temporal reasoning database) and FORBIN (the planner which uses TMM) (Dean 1985), and Elleby's TEMPO (Elleby 1986). Other important research in temporal reasoning and planning are NONLIN incorporated with the idea of *Clouds* (Tate 1986), and ISIS, a job shop scheduler which is built in with a vast amount of expert knowledge (Fox 1983,84).

Acknowledgements

I am grateful to Jim Doran for his supervision of this project and comments on this paper. Sam Steel and Richard Bartle have given me valuable opinions in both the content and the presentation of this paper. This project is supported by the studentship fund of the Department of Computer Science, University of Essex.

REFERENCES

Allen, J.F. & Koomen, J.A., 1983a, *Planning using a temporal world model*, IJCAI-83, 741-747

Allen, J.F., 1983b, *Maintaining Knowledge about Temporal Intervals*, CACM vol.26, no.11, 832-843

Allen, J.F. & Hayes, P.J., 1985, *A common-sense theory of time*, IJCAI-85, 528-531

Bell, C. & Tate, A., 1985, *Use and Justification of algorithms for managing temporal knowledge in O-plan*, AIAI-TR-6, University of Edinburgh

Dean, T., 1985, *Temporal Imagery: An Approach to Reasoning with Time for Planning and Problem Solving*, Ph D Dissertation, Yale University

Eleby, P., 1986, *In defence of point-based temporal reasoning*, Proceedings, 5th Alvey Planning SIG workshop

Fikes, R.E. & Nilsson, N., 1971, *STRIPS: a New Approach to the Application of Theorem Proving to Problem Solving*, AI-2

Fox, M.S., Allen, B.P., Smith, S.F. & Strohm, G.A., 1983, *ISIS : A Constraint-directed reasoning approach to job shop scheduling: system summary*, CMU Robotics Institute Technical Report CMU-RI-TR-83-8

Fox, M.S. & SMITH, S.F., 1984, *ISIS — A Knowledge-based system for factory scheduling*, Expert Systems, Vol.1 No.1, 1984, 25-49

Freuder, E.C., 1978, *Synthesizing constraint expressions*, CACM Vol 21, No 11, 958-966

Freuder, E.C., 1982, *A sufficient condition for backtrack-free search*, J ACM Vol.29 No.1, 24-32

Haralick R.M. & Elliott G.L., 1980, *Increasing tree search efficiency for constraint satisfaction problems*, AI 14, 263-313

Mackworth, A.K., 1977, *Consistency in networks or relations*, AI 8(1), 99-118

Mackworth, A.K. & Freuder, E.C., 1985, *The complexity of some polynomial consistency algorithms for constraint satisfaction problems*, AI 25, 65-74

Sacerdoti, E.D., 1974, *Planning in a hierarchy of abstraction spaces*, AI 5(2), 115-135

Sacerdoti, E.D., 1975, *The Nonlinear nature of Plans*, IJCAI-4, 206-214

Steel, S., 1985, *Refinments of operator-based action representation*, AISB-85, 98-107

Tate, A., 1976, *Project Planning using a Hierarchic Nonlinear Planner (NONLIN)*, D.A.I. Report No.25, Univ. of Edinburgh

Tate, A., 1977, *Generating project networks*, IJCAI-5, 888-893

Tate, A., 1984, *Goal Structure — Capturing the intent of plans*, European Conference of A.I. (ECAI) 84, 273-276

Tate, A., 1986, *Goal Structure, Holding Periods and "Clouds"*, Personal communication

Tsang, E.P.K., 1986a, *Plan Generation using a Temporal Frame*, ECAI-86

Tsang, E.P.K., 1986b, *The Interval Structure of Allen's Logic*, Technical Report CSCM-24

Tsang, E.P.K., 1987, *Planning in a temporal frame: a partial world description approach*, PhD thesis, to appear

Vere, S.A., 1983, *Planning in Time : windows and duration for activities and goals*, IEEE Transactions on Pattern Analysis and Machine Intelligence, Vol PAMI-5 No.3, 246-267

Vere, S.A., 1985a, *Splicing plans to achieve misordered goals*, IJCAI-85, 1016-1021

Vere, S.A., 1985b, *Temporal scope of assertions and window cutoff*, IJCAI-85, 1055-1059

Reasoning About the Physical World

PREDICTING THE BEHAVIOUR OF DYNAMIC SYSTEMS WITH QUALITATIVE VECTORS

A.J. Morgan
Computer Laboratory, University of Cambridge
Corn Exchange Street, Cambridge CB2 3QG, England
(Current address: Systems Designers Scientific, Pembroke
House, Pembroke Broadway, Camberley, Surrey, England)

ABSTRACT

Several classes of problem require the ability to reason about dynamically changing situations. The behaviour of an automated reasoning system will be determined in part by its internal representation of the dynamics of the situation. A method is discussed which describes dynamically changing relationships in terms of a sequence of qualitative vectors, rather than the more usual scalar representation. This allows operations such as differentiation and integration to be carried out on qualitative values. Examples are given of relationships expressed as sequences of qualitative vectors, including behaviours identified in closed–loop systems.

INTRODUCTION

Established quantitative approaches allow us to predict the behaviour of complex physical situations in a systematic way. For instance, the dynamics of linear systems can be analysed by several well developed methods; see (Palm 1983). However, there are circumstances in which quantitative approaches are not easy to apply; for example, where we have only limited information on the values of system variables.

Dynamic physical systems have been described by de Kleer, Brown, Kuipers, Williams and others, in terms of function and structure, and by Forbus in terms of physical processes (see references). These descriptions make use of the qualitative values of quantities and their first and higher–order derivatives, represented in terms of values drawn from the set $\{+0-?\}$, with ? representing an undetermined value. Many of the existing approaches use some technique of propagation of qualitative values through a network defining the system being analysed. This becomes difficult for realistic systems, particularly if they contain feedback loops. It is also difficult, or even impossible, to integrate (or differentiate) qualitative values, without having access to either analytic expressions or numeric values from which the qualitative integrals (or derivatives) can be derived. In (de Kleer & Brown 1984), for example, we find the statement: " ... $\delta n + 1x$ can not be obtained by differentiating δnx. One always has to go back to the quantitative definition ...".

This paper suggests a method of analysing dynamic systems which goes some way towards overcoming these difficulties. The examples of behaviour descriptions given later in the paper are drawn from basic electronics and mechanics, but the techniques are also generally applicable to thermal, or fluid physical systems, or indeed to other kinds of problem such as ecological systems (*e.g.* interactions between animal populations), management systems (*e.g.* control of warehouse stock levels), or financial systems (*e.g.* commodity brokerage).

CHANGES IN VALUE

Most real systems result in continuous quantities. "Continuous" in this context means that the quantities in the system are continuously differentiable (and have smooth rather than abrupt changes between values). In most work on qualitative reasoning, this intuition about physical systems is explicit – for example, see (Williams 1984) – or implicit. In certain types of system we may choose to ignore continuity. An example is in the domain of digital electronics, where we are often interested in states which persist for a much longer time than transition periods between them. For instance, a waveform may be assumed to be "square" if its rise and fall times are insignificant compared with other features in the system under consideration. However, a Fourier analysis shows that an infinite series of odd harmonics is required for truly square corners – a condition which is never achieved in reality.

This paper concentrates on the relationship between the *level* of a quantity and the *rate of change* of a quantity (or, equivalently, between a *level* of a quantity and the *time integral* of a quantity). An example is the relationship between the current through a capacitor and the voltage across its terminals. Assuming an ideal capacitor, the relationship between voltage and current is given by:
$$i_c = \frac{d}{dt}(v_c) \quad \text{or, equivalently,} \quad v_c = \int i_c \, dt + v_0$$
We can see that the voltage across the capacitor varies from its initial value of v_0 in a way that depends on the history of current flow. Past values of current flow can therefore be said to have caused the current value of voltage. This is sometimes referred to as "integral causality" (Palm 1983). There are well-known techniques for solving differential equations, but here we are more interested in reasoning about the behaviour of a system than in finding the value of a quantity at a particular instant.

The dynamic (*i.e.* time varying) behaviour of a system will be constrained by temporal relationships within it. In the case of systems which contain no temporal relationships (for example, a purely resistive network) there seems to be little point in considering rate of change or integral values as constraints on the system behaviour. This view differs from some other work on qualitative reasoning. For example, in (Williams 1984) a resistive network is analysed in terms of rates of change of current and voltage.

CURVE SHAPES

Before giving a method for describing variations in value, it is necessary to introduce the idea of a *curve shape*. If y is some continuously differentiable function of x, then repeated differentiation of y with respect to x will produce a series of values. The series given by the signs of the derivative values so obtained will be characteristic of the relationship between the variables involved. In graphical terms, the series of derivative signs will define a curve shape. The derivative signs can be represented by symbols, chosen from the set $\{+0-\}$ (with " ? " also used, to represent an undefined value). As examples; the curve shape for $y = (1-e^{-x})$ can be expressed as $[+-+-+ \ldots]$, the curve shape for $y = x^3$ as $[+++00 \ldots]$, and so on. For typographical convenience, in this paper the nth derivative of a quantity, y say, is shown generally as $\delta^n y$ and the curve shape as $[\delta^1 \, \delta^2 \, \delta^3 \ldots]$.

In practice, it is unrealistic to assume that we can always obtain a derivative series of sufficient length to give a completely accurate identification of the relationship between x and y. For example, we may not have measurements of the rate of change of a quantity, but may be forced to approximate by taking successive differences between adjacent values. Approximate differentiation is notoriously inaccurate; noise present in the measurements could introduce errors which would nullify an attempt at more accurate identification. A pragmatic approach is therefore to truncate the series at a point which gives a reasonable indication of the relationship between x and y. It appears, on empirical grounds, that sensible discrimination can be achieved by using only the first two qualitative values from the series (*i.e.* δ^1 and δ^2). Since there are three possible values for each member of the series, nine general relationships can be described. These are summarised pictorially in Figure 1. It is convenient to label these relationships with the pair of $\delta^1 \delta^2$ symbols which describe them; for example "++" indicates an increasing curve, concave up.

It is important to note that the curve shape gives information about the *form* of the relationship between y and x, rather than values. For example, both $y = x^{-2}$ and $y = e^{-x}$ can have the same curve shape, since the graph of y against x is similar in both cases (a decreasing curve with concave side up). In some cases extra information can distinguish between curves, but this is not always necessary.

QUALITATIVE VECTORS

To complete the qualitative description of a varying parameter we also
need the 0th derivative, *i.e.* the qualitative value of the level of the
quantity. This can be combined with the curve shape to form a *qualitative
vector* made up of the [δ0δ1δ2] triplet. The general qualitative vector
identifies the relationship between two quantities over a region in the x,y
plane. Assume that we wish to find a qualitative vector for y in terms of x
(we can note in passing that it may not always be possible to find a unique
inverse relationship: *i.e.* x in terms of y). A convenient way of expressing
the relationship is to give a y vector for each of a series of points or
intervals on x. In this paper, "points" on x are treated as intervals of
infinitesimal width. Continuity is ensured by rules which limit possible
sequences of qualitative vectors, as discussed in the next section

The convention adopted here is that the vector sequence describes y values
for ascending values of x. The sequence itself only gives an ordering rather
than a value to the vectors. In some cases we may be able to assign x
values to the vectors. It may also be possible to assign y values to the
vectors. In either case the values may be crisp or fuzzy. It is possible to
reason about relationships even if neither x nor y values are available. In
the vector sequence: [?00], [?++], [?+0], [?+−], [?00], for
example, we can identify a region of linear operation, in which y is
proportional to x (*i.e.* [?+0]), and can also observe that beyond a certain
point the increase or decrease in y with x slows down and finally stops.
This is characteristic of a saturation curve, such as might be found in an
amplifier. The recognition of such features is achieved without needing to
specify values or intervals for x or y. Of course, additional information
allows improved reasoning. In the above example, we may have a
specification for the region of linear operation, which would allow some
prediction to be made about the changes in y over a range of values in x.
An example of a relationship which involves a sequence of qualitative
vectors is the rising voltage edge given in Figure 2.

A qualitative vector arithmetic can be defined as a straightforward
extension to the scalar case, much as numerical vector operations are
related to scalar numerical operations. As examples; multiplying a
qualitative vector by a constant multiplies each vector element by the
constant; vector addition sums the vector elements involved; and so on.

CONTINUITY

Considering the nine possible curve shapes in Figure 1, we can see that
certain sequences could not exist in reality. For example, in a
time–varying relationship,"?+0" could not be followed immediately by
"?−0" without some intervening episodes (however short) in which the

curve gradually changed from a positive to a negative slope. Of course, the length of the intervening episodes may be insignificant when compared with other episodes, and possibly even be ignored completely when reasoning about major effects. However, it is necessary to ensure that relationships are correctly represented in a competent reasoning system so that the system is not caught out by having a model which is too simple to provide an adequate representation of its domain.

The rules governing possible sequences of qualitative vectors can be expressed in terms of pairs of vectors: using predecessor and successor vectors, each of which can be assumed to have three elements; $[\delta^0 \delta^1 \delta^2]$. The rules define the vector sequences which can represent physically valid situations. Most previous work on qualitative reasoning makes use of similar rules.

Continuity Rule 1
Elements can only change to the opposite sign by a transition through zero.

Rule 1 prevents any of the elements of the qualitative vector changing directly from "+" to "−" or from "−" to "+" in successive states.

Continuity Rule 2
If any element δ^n has the same value in the succeeding vector δ^n_{succ}, its lower order element must behave consistently, as defined by:
$$\delta^{n-1}_{\text{succ}} = 0 \text{ or } \delta^{n-1}, \text{ if } \delta^{n-1} = (\text{minus } \delta^n)$$
$$\delta^{n-1}_{\text{succ}} = \delta^{n-1} + \delta^n, \text{ otherwise}$$

Rule 2 uses the fact that each element is the rate of change of its lower order element (except for the first element). It follows that a low order element cannot increase if its higher order element defines a decrease (and vice versa) nor change in any direction if its higher order element defines it as stationary.

Continuity Rule 3
Any vector of the form $[? \delta^1 \delta^2]$ where $\delta^1 = 0$ will be <u>preceded</u> by a vector of the form $[? \delta^1_{\text{pred}} \delta^2_{\text{pred}}]$ and <u>succeeded</u> by a vector of the form
$[? \delta^1_{\text{succ}} \delta^2_{\text{succ}}]$ with the following constraints:
$$\delta^1_{\text{pred}} = (\text{minus } \delta^2_{\text{pred}})$$
$$\delta^1_{\text{succ}} = \delta^2_{\text{succ}},$$
    and, additionally, if $\delta^2 \neq 0$ then
$$\delta^1_{\text{pred}} = (\text{minus } \delta^2)$$
$$\delta^1_{\text{succ}} = \delta^2.$$

Rule 3 ensures that "flat" parts of relationships (where the first derivative is zero) are inter-connected by curves of the correct shape.

Continuity Rule 4
Any vector with $\delta n = 0, \delta n + 1 \neq 0$ will represent an infinitesimal interval in x.

For temporal relationships such vectors can be considered point–like. The short duration of the intervals means that they cannot be associated with stable system states. Their role is to interlink states of significant duration in a way that maintains continuity. Rule 4 follows from the underlying assumption that "zero" is a singularity.

One immediate advantage of the rules is a reduction in the space of possible "next states". Considering all possible combinations of sign would produce a large number of potential state transitions. Even with the truncated curve shapes used here (which give 3–element qualitative vectors), a simplistic view gives $3^6 = 729$ possible transitions. Applying the rules above to limit the transitions to feasible cases results in a dramatically reduced range of possibilities, which can be shown in the form of a directed graph. Figure 3 shows the effect of continuity rules 1 to 3 in terms of feasible transitions between adjacent curve shapes, using a 3–element vector.

HIDING DETAIL

If we are reasoning about a large system, it may be convenient to remove some of the detail in certain areas so that we can concentrate effort where it is most needed.

Compressing vector sequences. We can make use of the feasible transition graph to compress a sequence of qualitative vectors without introducing ambiguity. This allows a compact expression of the relationship (which can be expanded, if necessary, for more detail) and, more importantly, avoids the need to propagate every state change within a system.

The feasible transition graph can be used to define minimum transitions. The minimum transition between any two curve shapes is the shortest path between $\delta 1 \delta 2$ values on the transition graph which is consistent with the corresponding $\delta 0$ values. For example, assume that the rising voltage edge in Figure 2 had been described by the compressed sequence: $[\,000\,]$, $[\,+00\,]$ (corresponding to values of V before t_0 and after t_6). The value of $\delta 0$ has changed from "0" to "+" over the sequence, implying that at some time between t_0 and t_6 the value of $\delta 1$ must have been "+". The minimum transition between $\delta 1 \delta 2 = "00"$ and $\delta 1 \delta 2 = "00"$ which includes δ^1 = "+" is the $\delta^1\delta^2$ sequence "00", "0+", "++", "+0", "+−", "0−", "00", which corresponds to the sequence shown in Figure 2.

Compression of vector sequences is possible providing that compression can be achieved without ambiguity, *i.e.* there is a unique minimum transition between the compression end values. This is possible in the case of Figure 2, but would not have been possible had the initial value of δ^0 been "+" instead of "0". In this case, without a change in δ^0, it would not be possible to distinguish between two possible transitions, one passing through δ^1 = "+" and the other passing through δ^1 = "−". The ambiguity arises from the result of the qualitative sum of opposite values being undefined.

Expanding vector sequences. A relationship may be expressed in a form which requires expansion before a reasoning process can make use of it. Compressed sequences could be produced by a previous compression operation, or by some operations on qualitative vectors which generate compressed sequences. Expansion of a vector sequence assumes that the $\delta^1\delta^2$ values of the intervening vectors are minimum transitions, and that δ^1 behaves consistently. For example, the vector sequence [+++], [++−] is not feasible (violating Continuity Rule 1), but can be made feasible by expansion to [+++], [++0], [++−]. Although minimum transitions can define the $\delta^1\delta^2$ values, it may not be possible to determine the δ^0 value without further information. For example, the vector sequence [−++], [++−] expands to [−++], [?+0], [++−], and may expand further when additional information identifies the exact sequence between the extreme vectors. Again, this ambiguity is a consequence of combining qualitative values of opposite sign.

Expansion of a vector sequence by repeating vectors is possible, but not normally useful, since we can always express the extended range of the y vector by defining a suitable x interval.

QUALITATIVE CALCULUS

So far a relationship between x and y has been discussed in terms of a sequence of y qualitative vectors for points or intervals in x. However, it is also possible to use very straightforward operations to obtain new qualitative vectors which are the integral and the derivative of y with respect to x. The results of the integration and differentiation operations are new qualitative vectors composed of the usual [$\delta^0\delta^1\delta^2$] elements, with values chosen from the {+0−?} set. Only the integration rule is required for the examples given in this paper, but the corresponding differentiation rule can be expressed in a similar way.

The only complexity in integration arises in determining the δ^0 value of the integral. Just as in the more conventional mathematical counterpart, we need to take into account a constant of integration to form the δ^0 element. This may introduce the usual ambiguity due to qualitative summation, which must be resolved with other information.

> *Integration Rule*
> If the vector for y is made up of the elements $\delta^0_y \delta^1_y \delta^2_y$, and its integral by the elements $\delta^0_I \delta^1_I \delta^2_I$, then
> $$\delta^2_I = \delta^1_y$$
> $$\delta^1_I = \delta^0_y$$
> δ^0_I is determined by additional information, including initial conditions.

In effect this shifts the vector one place to the right, shifting in a new value for δ^0 and truncating the integral to retain the consistent 3–element vector. When the integration operation is applied to a sequence of vectors, it may generate an "expanded" sequence which includes repeated vectors. This is an effect caused by the limitation of the qualitative vector to three elements, which can lose the fine distinction between certain adjacent vectors. A simple example can be seen by considering the integral of the rising voltage edge in Figure 2, with δ^0 defined by assuming zero initial conditions.

EXAMPLES

The qualitative vector approach outlined above can be used in several applications. Two examples are given here of envisioning the behaviour of simple physical systems. The general method can be summarised in the following steps.

(1). Define a model which adequately represents the system.
(2). Develop equations which express the relationships in the model.
(3). Define the initial conditions and driving functions as qualitative vectors.
(4). Use the equations from (2) and the integration rule to generate all qualitative vectors (for each variable in the system) which are simultaneously consistent with each other.
(5). Use the continuity rules to establish the ordering of the resulting vector set.
(6). Filter the results, if required.

Step 1 is not required here, as the examples are drawn from previous literature. Both are examples of linear systems; one first order and the other second order. The system equations of step 2 can be developed in a systematic way, which is not explained here, but can be found in standard textbooks on system dynamics. The initial conditions in step 3 use the same values used in the referenced papers (although the method will accept any initial values).

Example 1. In (Williams 1984), Temporal Qualitative (TQ) analysis is used to investigate the relationship between voltages and currents in an electrical R-C circuit. The circuit and a corresponding block diagram are shown in Figure 4. This is a linear system, with governing equations:

$$v_R = i_R.R \quad v_C = \frac{1}{C}\int i_C \, dt. + v_0$$
$$v_R = v_C \qquad i = i_C + i_R$$

The initial conditions are the same as in (Williams 1984): $v_C = 0$, and a step current input is applied. In qualitative vector terms, the driving function is $i = [+00]$, and the initial conditions are $i_R = [000]$, $i_C = [000]$.

The values of both C and R are positive constants, so i_R will be the qualitative integral of i_C. This means that $\delta^1\delta^2$ in i_R are constrained to be the same as $\delta^0\delta^1$ in i_C. The other constraint arises from the system equations and the driving function, which together limit the sum of i_C and i_R to be the constant value $[+00]$.

The complete set of vectors consistent with these constraints is as follows:

state	i_R	i_C
1	[++−]	[+−+]
2	[+00]	[000]
3	[+−+]	[−+−]
4	[0+−]	[+−+]
5	[−+−]	[+−+]

The constraints are applied in a way which avoids generating large numbers of alternative qualitative values. For example, in state 1 the vector sum of i_R and i_C must be $[+00]$ (as it is in every state). If we add the values of iR and iC we get a vector $[+??]$. This has nine possible interpretations, but we do not have to handle all of them; merely note that $[+00]$ is among the possibilities, and therefore consistent with the constraints.

The steady state (at $t = \infty$) can be found easily. At steady state all transients will be zero. Since i_C is the derivative of i_R it must be zero, and so from the system equations and the defined driving function $[+00] = [000] + i_R$ at $t = \infty$. The only possible value of i_R which satisfies this equation is $[+00]$, which corresponds to state 2. The remaining states represent responses to different initial conditions. Since we have information on the initial conditions we can filter the set of states to form the subset required. In state 1 both currents are in the directions shown in Figure 4(a), with the capacitor current falling and the resistor current rising. In state 3 the capacitor current is reversed, corresponding to an initially high positive charge on the capacitor, which is discharging through R. In state 5 the resistor current is reversed, corresponding to a negative initial charge on C. State 1 therefore represents the persistent conditions while i = [+00].

There are a number of points which should be noted. Firstly, a complete history of values tracing the changes in current from the moment i is applied is not necessarily generated. This is because we are considering the responses of the system during a period when i is steady at [+00]. If, instead of assuming that i changes from [000] to [+00] we take the expanded sequence for i (as in Figure 2) we can generate a full history of states for iC and iR. This is an example of "hiding" detail which is not required. We do not have to generate information about intermediate states to arrive at the one we want. A second important point is that there is no sequence leading to the steady state. This agrees with quantitative systems theory, in which the steady state is approached asymptotically, only being achieved at $t = \infty$. Thirdly, it is worth pointing out that the behaviours predicted by this method agree with those from quantitative techniques, within the limits of the coarse quantisation represented by the qualitative vectors. For example, in the case considered above, the capacitor current i_C is equal to (i.e. $e^{-t/CR}$). The signs of the first and second derivatives of this expression are "−+", agreeing with the values found in state 1.

Example 2. In (de Kleer & Bobrow 1984), de Kleer and Bobrow discuss the pressure regulator example, and in particular the dynamic behaviour of the mass–spring–damper subsystem which can be identified within the regulator. As in (de Kleer & Bobrow 1984), the treatment here is limited to the mechanical subsystem within the regulator. Again, this is a linear system, which can be described by the following equations:

$$f_d = B.v_d \qquad f_s = K(\int v_s \, dt. + v_0)$$
$$f_m = M.d/dt \, v_m \qquad v_m = v_d$$
$$v_m = v_f \qquad F = f_m + f_s + f_d$$

where v = velocity, f = force; subscripts d, s, and m refer to the damper, spring, and mass respectively; and B, K, and M are positive constants which define the damper, spring, and mass respectively. A physical schematic and a block diagram are shown in Figure 5. Although not clearly stated in (de Kleer & Bobrow 1984), the analysis of the system appears to assume a step force input which drives the regulator away from some static operating position (i.e. zero initial conditions).

As in the previous example, we can find a steady–state solution in which a = [000], v = [000] and x = [+00]. This agrees with intuitions about the action of the regulator; the force will move the diaphragm to a new (offset) equilibrium position. Of more interest is the dynamic behaviour of the system. To save space here we will assume that the displacement will always be positive. This implies that all values of x must have $\delta 0 = +$. The possible vectors which meet the constraints are as follows:

state	x	v	a
1	[+++]	[++−]	[+−?]
2	[++0]	[+0−]	[0−+]
3	[++−]	[+−+]	[−+?]
4	[++−]	[+−0]	[−0+]
5	[++−]	[+−−]	[−−+]
6	[+0+]	[0+−]	[+−?]
7	[+00]	[000]	[000]
8	[+0−]	[0−+]	[−+?]
9	[+−+]	[−++]	[++−]
10	[+−+]	[−+0]	[+0−]
11	[+−+]	[−+−]	[+−?]
12	[+−0]	[−0+]	[0+−]
13	[+−−]	[−−+]	[−+?]

Considering only the combinations of values of x and v, we can identify 13 different states. The same number of states is also identified in (de Kleer & Bobrow 1984), although the interpretation here is somewhat different.

States 2,4,6,8,10, and 12 are momentary states (from Continuity rule 4). State 7 corresponds to the steady state value. States 1,2,5,4,3,8,13,12,9,10,11,6,1, ... form a repeating cycle (by considering feasible transitions), which is illustrated in Figure 6. If values of the acceleration vector are taken into account, the qualitative analysis can also identify behaviours associated with different degrees of damping. The possibility of an oscillatory response is identified in (de Kleer & Bobrow 1984). However, there is a second possibility which is not identified in that analysis. Depending on the combination of values of M, K, and B, the response of the system may be underdamped, critically damped, or overdamped. The first of these possibilities gives rise to an oscillatory response as described above. The second two possibilities produce a response which converges towards the final value without oscillation. State 3 is an example of such a case. Suitable filtering of the set of states can extract the appropriate oscillatory or damped cases.

CONCLUSION

The use of constraint techniques and transition analysis appears to offer a powerful yet economic method of analysing industrial–scale dynamic systems. It is encouraging that all possible modes of response of a system can be identified. Normally, additional information can be used to establish which of several alternative behaviours will actually take place. For example, we may have approximate information about parameter values which will indicate whether an oscillatory mode of operation is possible.

Although outside the scope of this paper, it is worth pointing out that the form of the models used here to support qualitative descriptions, are very similar to the "block diagram" models used in conventional systems engineering. This feature adds flexibility, allowing any mix of quantitative or qualitative descriptions to be generated, depending on the needs of the requirement.

The approach outlined here is being used as part of a system to analyse the operation of an experimental petro–chemical plant. Possible future extensions include non–linear as well as linear components.

ACKNOWLEDGEMENTS

This work has been supported by Systems Designers Limited. Thanks are due to BP Research, Sunbury, for providing challenging problems, and to William Clocksin for guiding the research in appropriate directions.

REFERENCES

(de Kleer & Brown 1984)
de Kleer J. and Brown J. S. *"A Qualitative Physics Based on Confluences"*, Artificial Intelligence 24 (1–3), pp 7–83, 1984

(de Kleer & Bobrow 1984)
de Kleer J., and Bobrow D. *"Qualitative Reasoning with Higher–Order Derivatives"*, Proceedings of AAAI–84, pp 86–91, 1984

(Forbus 1984)
Forbus K. *"Qualitative Process Theory"*, Artificial Intelligence 24 (1–3), pp 85–168, 1984

(Kuipers 1985)
Kuipers B. *"The limits of qualitative simulation"* Proceedings of the Ninth International Joint Conference on Artificial Intelligence, pp 128–136, 1985.

(Palm 1983)
Palm W.J. *"Modeling, Analysis and Control of Dynamic Systems"*, John Wiley & Sons, 1983

(Williams 1984)
Williams B.C. *"Qualitative Analysis of MOS Circuits"*, Artificial Intelligence 24 (1–3), pp 281–346, 1984

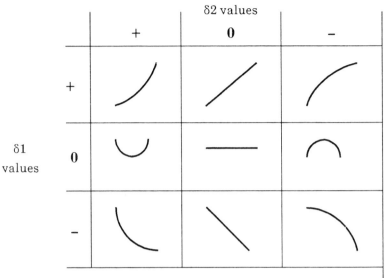

Figure 1. Possible $\delta_1 \delta_2$ combinations

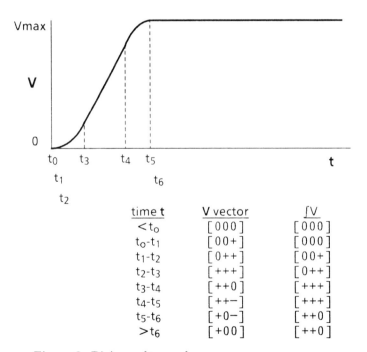

time t	V vector	∫V
$< t_0$	[0 0 0]	[0 0 0]
t_0-t_1	[0 0 +]	[0 0 0]
t_1-t_2	[0 + +]	[0 0 +]
t_2-t_3	[+ + +]	[0 + +]
t_3-t_4	[+ + 0]	[+ + +]
t_4-t_5	[+ + −]	[+ + +]
t_5-t_6	[+ 0 −]	[+ + 0]
$> t_6$	[+ 0 0]	[+ + 0]

Figure 2. Rising voltage edge

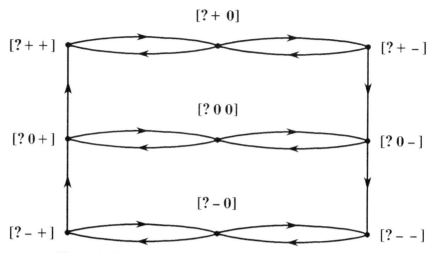

Figure 3. Feasible transition graph

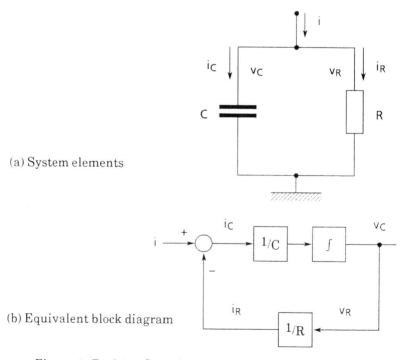

(a) System elements

(b) Equivalent block diagram

Figure 4. Resistor-Capacitor system

(a) System elements

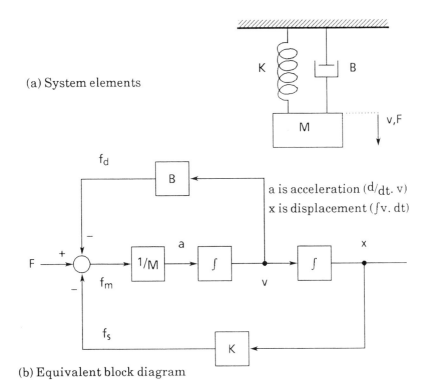

a is acceleration $(d/dt. v)$
x is displacement $(\int v. dt)$

(b) Equivalent block diagram

Figure 5. Mass-Spring-Damper system

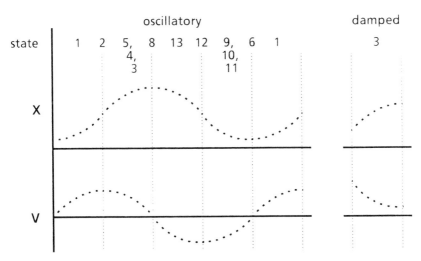

Figure 6. Predicted Behaviour for Mass-Spring-Damper

COMMONSENSE REASONING ABOUT FLEXIBLE OBJECTS: A CASE STUDY

Mauro di Manzo* and Emanuele Trucco**

ABSTRACT

Intelligent autonomous robots must be able to reason about entities of the real world. Many applicative tasks involve flexible objects such as electric cables, paper sheets and ropes, but reasoning about nonrigid objects is quite a hard computational task. While numerical simulation is not viable in this case, qualitative reasoning seems to provide a very promising approach. We present in this paper a study regarding the physics of the domain of one–dimensional flexible objects such as strings. The representation of situations and changes is provided according to the formalism of Qualitative Process Theory (QPT). The model aims to provide a formal description of the domain dynamics as complete as possible, allowing a recognition and consistency analysis of the active processes, and to enable a reasoning system to anticipate the evolution of a given situation, possibly for planning inteventions towards the achievement of a given goal. The elements of the model, their function and use are introduced and discussed. An example of dynamic situation with strings evolving through different states is given.

Introduction: the rubs of strings

"Cognitive science is sometimes reminescent of alchemy"
P.Hayes, the Second Naive Physics Manifesto

Intelligent autonomous robots moving about in the world must be able to understand a given situation, predict its evolution and modify it by performing some planned action. Many tasks relevant to real applicative frameworks, such as placing electric cables for instance, involve flexible things which hardly undergo the idealizations holding for rigid objects. In spite of the computational difficulties arising in the attempt of modeling them, human beings are quite good at manipulating strings on the basis of incomplete, qualitative information. That's why qualitative physics seems to provide the most promising approach to develop models apt to cope with the above mentioned problem.

An exhaustive qualitative analysis of the string world is still lacking. More attention seems to have been devoted to liquids [1] and the understanding of engineered devices [2]. Forbus [3] gives a short account for elastic flexible objects introducing the notion of push– and pull–transmitter and extends it for describing the behavior of strings. However, his proposal accounts only for a restricted number of situations, excluding e.g. the case of a rope hanging over a pulley. He states also that a full description would also require tension, and that's exactly one point that we have tried to meet.

---

* Institute of Computer Science, Univ. of Ancona, via Brecce Bianche, 60100 Ancona, Italia.

** Departement of Communication, Computer and System Science, via all'Opera Pia 11/a, 16145 Genova, Italia.

In a quite different approach, Gardin et al. [4] moves from the central idea of a fundamental molecular ontology. They propose an analogical representation for nonrigid objects where each "molecule" knows about its local properties and possible interactions with its neighbours. While this approach appears to have a strong point in "naive simulation", it is not clear how global qualitative inferences could be drawn without having the situation actually evolve using an analogical representation.

De Kleer and Brown [2] outline that the behavior of complex systems arise mostly from structure. In their confluence−based approach every physical situation is regarded as some type of physical device made up of individual components, each of which contributing to the behavior of the overall device. This results in a very mechanistic world view, well−suited for those situations where decomposition in functional subpart is appalling, e.g. analysing engineered devices. But the structural decomposition is at a loss at coping with objects for which neither structural nor functional subpart can be definitely individuated.

This paper proposes a qualitative model for the dynamics of one−dimensional flexible objects. The ontology used is that of QPT and is made of objects, states and processes. The next section is a short review of some concepts of QPT. Subsequently, some features of the proposed model are discussed. Then we introduce examples of the elements we have defined and give an example of dynamic modelling.

The Qualitative Process Theory

This section revises quite briefly the main concepts of Forbus's QPT relevant to our model. The interested reader is referred to [3, 5].

QPT extends the ontology of common sense physical models by introducing the notion of **physical process**. Processes are an extension of Hayes's histories [1] in that they help a) individuate those intersections of histories which correspond to interactions between objects and b) account for the generation of histories themselves. A collection of processes describes the evolution of any situation, that is how things change in time. Objects properties and states can change dramatically: heated water can start to boil, a string breaks under the action of too great a force, and so forth. Different and meaningful states of objects need therefore to be represented in order to have processes recognize or predict changes. This is accomplished by **individual views** (IV). IV's are composed of objects and specify their state through a representation of their properties called **quantities**. Each quantity consists of an **amount** and a **derivative**; both of them are qualitative descriptions of numbers. Relationships among quantities are modeled in terms of functional dependency by means of a **qualitative proportionality** operator (α_Q). Both processes and IV's become active only if their **preconditions** and **quantity conditions** are satisfied. These account respectively for physical and commonsense preconditions, meaning to say that water needs heat to boil (physical precondition) but also someone placing the pot on a stove (commonsense precondition or quantity condition).

Most importantly, the assumption is made that all changes in physical systems are caused directly or indirectly by processes. Their direct effects are stated explicitly in an **influences** field. Their indirect effects are propagated via the qualitative proportionality operator. As a consequence, the physics for a domain must include a **process vocabulary** that occur in that domain. Thus the qualitative description of a given situation is composed of a collection of objects, their state expressed by their quantities in the IV's, the relations holding among quantities, and the processes active.

The last kind of QPT structure worth mentioning are the so−called **encapsulated histories** (EH's). Unlike processes, they explicitly refer to a particular sequence of times during which changes take place. They describe easily such phenomena as collisions between moving objects, whose description in terms of processes would be complicated or unclear.

Description and motion: are strings aristotelian?

A commonsense system which a robot could actually use to interact with its environment must cope with two main problems:

1) provide a formal language for describing a given situation in terms of its elements. On such a description both a detection of active processes and a consistency analysis can take place. These two categories of inferences are known as **activity determination** and **skeptical analysis**.

2) how to plan actions aiming to change the evolution of a given situation. This point requires a qualitative dynamic model capable of deducing what is going to happen given some formal description, what Forbus has called the **prediction problem**.

The first point has been widely treated by Forbus [3]. The QPT ontology has been used here to encompass the knowledge necessary to model the dynamics of the strings domain. The next section reviews the processes, IV's and EH's we have defined and gives some details.

The second problem leads us to an interesting approximation which turns out to be reasonable and powerful for strings. We call it the **aristotelian motion hypothesis**. It is suggested by the behavior of strings hanging or laying on surfaces in the real world. Let's start stating that "hanging" and "laying" can be reasonably regarded as the only two possible "states" for a string. Strings suspended by one end but partially laying can be thought of as two simple strings exchanging tension in a shared point. Now, hanging strings tend to assume a catenary shape in any case, under the action of the gravity force. In this case the aristotelian motion hypothesis is trivial, because the gravity force never ceases its action. Laying strings, conversely, are mostly subject to friction and their motion on a real plane hardly keeps on after the applied force dies. Tilted surfaces causes a behavior which is simply a fall along the surface; or it can qualitatively be assimilated to that of a hanging string when the tilt can overcome frictions.

This approximation let us cut out a good deal of worries about strings dynamics. The conclusion is that a string or a system of strings can move under the following two conditions:

1) there is a force applied and its intensity is greater than a "resistance" depending on friction and weight. This threshold is part of the information held by IV's.

2) the resistance of obstacles is added to that of the moving string. This implies the existance of an event which models the impact with ground and with constrained or heavy objects. An encapsulated history, IMPACT, accomplishes this function.

The objects of the string world

There we come to the elements of our qualitative model of strings behavior. This section introduces the representation of the physics domain, the relations between quantities and the process vocabulary used. An example of their use is given in the next paragraph.

The first primitive object of the world is of course the **string**. Strings are one—dimensional and can change their shape and position according to the presence of other objects and the forces acting on them. The definition for a string is given in figure 1 using logical notation.

This description gathers intrinsic properties (thickness, length, flexibility, break force, etc.) together with quantities regarding spatial information. In particular, **left(s)** and **right(s)** give the coordinates of the extreme points in a fixed reference frame. They are actually triples of numbers. The quantity **extr__points__distance** is the distance between the two ends of the strings. **Inclination** and **direction** are calculated only on the basis of the endpoints position. They give rough information about the string position with respect to the horizontal and vertical plane in the reference frame. These two quantities can assume discrete values as indicated in fig.2.

Finally, **projection** individuates the shape of the string projection on the horizontal plane. The last two predicates in fig.1 express the fact that strings that are not sustained must lay on the ground.

It might be necessary for some tasks to know precisely the position of a segment of string. In this case the concept of **string unit** is introduced, whose representation follows:

u ∈ string__unit < − > u ∈ string AND length(u) = 1

While for a generic string the orientation and direction quantities are qualitative representation of numbers, they are defined without ambiguity for such string units.

The second element of our world is the **support**. Its relevant intrinsic properties are position, size and the maximum tension which it can bear. Another interesting fact is that a hanging string can present a discontinuity at the supporting point. This is represented in figure 3. Do notice that the definition given allows a string to be a support.

STRING(s)
s ∈ STRING < − >

Has__quantity(s,length)	AND	Has__quantity(s,thickness)	AND
Has__quantity(s,flexibility__coeff)	AND	Has__quantity(s,friction__coeff)	AND
Has__quantity(s,break__force)	AND	Has__quantity(s,unit__weigth)	AND
Has__quantity(s,left__end)	AND	Has__quantity(s,right__end)	AND
Has__quantity(s,extr__points__distance)	AND	Has__quantity(s,min__height)	AND

Has__quantity(s,inclination) AND Has__quantity(s,direction) AND Has__quantity(s,projection) AND (end(s) = left(s) OR end(s) = right(s)) AND (NOT(phys__contin(s,ground) − > EXIST(o) : support(o) AND phys__contin(s,o))

Figure 1

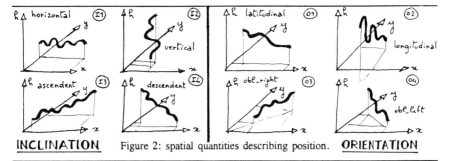

INCLINATION   Figure 2: spatial quantities describing position.   **ORIENTATION**

SUPPORT(s)
s ∈ SUPPORT < - >

Has_quantity(s,length)   AND   Has_quantity(s,thickness)   AND   Has_quantity(s,depth)
AND   Has_quantity(s,position)   AND   Has_quantity(s,max_tension)   AND
Has_quantity(s,weight) AND (∀ s1,s2 C STRING_UNIT : phys_contin(s1,s) AND
phys_contin(s2,s)   - >   (NOT(orientation(s1)=orientation(s2))   OR
(NOT(inclination(s1)=inclination(s2)))

Figure 3: the description of the object class SUPPORT.

The last basic element is the **force**. It is defined simply by

FORCE(f)
f ∈ FORCE < - > Has_quantity(f,intensity) AND Has_quantity(f,direction).

For the time being we consider only horizontal and vertical forces.

States, relations and processes

String properties can change in time. Some of these changes depend on values of quantities, e.g. if a catenary is gently let down, when its min_height point (see figure 1) reaches the ground we can consider it gone. Such states and changes are modelled with individual views. An example of IV is exactly the catenary whose description is given below. Further explanation is presented in graphical form in figure 4.

INDIVIDUAL VIEW catenary(c)

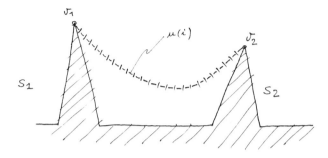

Figure 4: a catenary as composition of STRING__UNITS.

INDIVIDUALS:
u(0)....u(n) a STRING   /* made of string__units */
s(1), s(2) a SUPPORT

PRECONDITIONS:
∃k: (∀i) [(i<k) −> descendent(u(i)) AND inclination(u(i+1))<inclination(u(i))] AND
(∀j) [(j>k) −> ascendent(u(j)) AND (inclination(u(j+1))>inclination(u(j))];
phys__contin(u(0),s(1)) AND phys__cont(u(n),s(2));
(∀i) linked(u(i), u(i+1)).

QUANTITY CONDITIONS
$\sum_i$length(u(i)) > distance(s(1), s(2));
min[h(u(0)),.....h(u(n)) > 0.

RELATIONS
xyhdx(c) =< xyh(s1);
xyhsx(c) =< xyh(s2);
length(c) =<   length(u(i));
dist__estr(c)∝$_d$distance(s(1), s(2));
hmin(c)∝$_Q$+ dist__estr(c);
hmin(c)∝$_Q$- length(c);
curvature(c)∝$_Q$+length(c);
curvature(c)∝$_Q$-dist__estr(c);
orientation(c)∝$_Q$+orientation(s(1), s(2));
tension(c)∝$_Q$+length(c);
F1, F2 is FORCE
    horizontal(F1) AND appl(F1,c) AND verse(F1,s1,c);
    horizontal(F2) AND appl(F2,c) AND verse(F2,s2,c);

intensity(F2) = intensity(F1);
intensity(F1)$\propto_{Q^+}$tension(c).

The above description actually contains more information about thickness, friction, breakpoint and so forth, which have been omitted for conciseness. Similar IV's are defined for the catenary: VERTICAL_CAT(c) and WEIGHTING_CAT(c).
Another peculiar state is that of a taut string, modelled by the IV TAUT(c). Relation between quantities are represented in IV using the   operator. In the IV CATENARY, for instance, it is stated that the curvature of a catenary increases qualitatively with its linear length and decreases when the endpoints are brought nearer to each other.

The changes between different states are often due to phenomena like collision, e.g. an impact with the ground which alters the shape of a catenary. Such events are best modelled with descriptions which explicitly refer to time, the encapsulated histories. An example is the EH IMPACT:

ENCAPSULATED HISTORY impact(O1,O2)

INDIVIDUALS:
    E a TIME_INTERVAL
    O1, O2 a OBJECT

PRECONDITIONS:
    (T active (motion(O1, O2)) start(E))
    (T phys_contin(O1, O2) start(E))

RELATIONS:
    (M resistance(O1) during(E)) = (M resistance(O1) start(E) +
        (M resistance(O2) start(E))

Notice that the relation field refers explicitly to time, stating that the resistance of O1 during the interval E equals the sum of the resistances of O1 and O2 at the beginning of E.

The unique responsible for changes in the QPT ontology is the process. A string laid on a support can slide under the action of the gravity force, thus activating a process SLIDE; a support can be moved to modify a catenary, and so forth. Like IV's, processes have preconditions and quantity conditions through which it is possible both to detect changes in a situation (activity determination) and to plan actions aiming to reach a desired state. As an example, we introduce below the definition of the process SLIDE.

PROCESS slide(f1, f2)

INDIVIDUALS
    F1, F2 a FORCE
    f1, f2 a STRING
    t0,..., tn a STRING   /* optional */

PRECONDITIONS
    horizontal(F1) AND horizontal(F2)
    verse(F1, f2, f1) AND verse(F2, f1, F2)

applied(F1, t0) AND applied(F1, tn)
linked(f1, t0) AND linked(tn, f2)
(∀ i) (linked(t(i), t(i+1)) AND taut(t(i)))

QUANTITY CONDITIONS
intensity(F1) < intensity(F2)

RELATIONS
slide_speed $\propto_{Q+}$ (intensity(F2) − intensity(F1))

INFLUENCES
I+ [length(f2), slide_speed]
I− [length(f1), slide_speed]

It is interesting to notice that SLIDE has effects on the quantities length, curvature, tension, force applied at supports and min_height. The simultaneous change of this set of quantities let us recognize that a SLIDE process is active. According to the aristotelian motion hypothesis, SLIDE will terminate when the quantity condition
intensity(F1) < intensity(F2)

become false. The structure of SLIDE tells us that this can be accomplished only by increasing the intensity of F1. The IV's defined for string states reveal in turn that the only way to augment the force acting on a string is to apply a second force, which can be got e.g. by activating a process SUPPORT_MOVEMENT. Notice that this can be regarded as a simple chunck of some plan deduced through the model.

Example: the sliding string

Let's see now how the process SLIDE can model the real movement of a string sliding on its supports. Consider the situation in figure 5. A string is laid on four supports whose

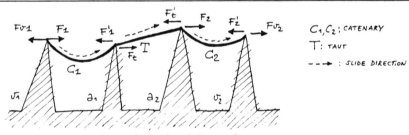

Figure 5: a string sliding on four supports.

sections are visible. Given an interval of time I, the effects of the evolution can be expressed by

(M length(C1) start(i)) < (M length(C1) end(i))
(M length(C2) start(i)) > (M length(C2) end(i))

The process responsible for that is SLIDE. For the forces implied by the IV's C1, C2, T (see figure 5) hold that

F1 = F'1 = F't (from the IV's C1 and T)
Ft = F'2 = F'2 (from the IV's C2 and T)

furthermore, being v1 and v2 constraints

Fv1 = F1   F'2 = Fv2

being the process SLIDE(C1,C2) active, we get F2 > F'1 and in short

Fv1 = F1 = F'1 = F't < Ft = F2 = F'2 = Fv2

From this situation the predictions listed in table 1 can be deduced.

Some events might make the process stop. These are modelled by the following EH's:
1) if min_height(C2) = 0, the EH CATENARY_TO_HANGING is activated (figure 6a). This results in the disappearance of C2 and consequent constitution of the segments f1, f2.
2) if curvature(C1) = ZERO the EH CATENARY_TO_TAUT is activated. In this case (figure 6b) C1 disappears and T1 is created in its place.

Now suppose that we want to stop the fall of C2 before SLIDE comes to one of its natural ends. Incidentally this implies that a process detection has already taken place. During

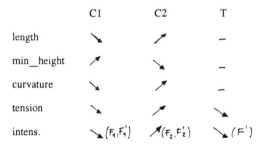

	C1	C2	T
length	↘	↗	—
min_height	↗	↘	—
curvature	↘	↗	—
tension	↘	↗	↘
intens.	↘ (F_1, F_1')	↗ (F_2, F_2')	↘ (F')

Table 1

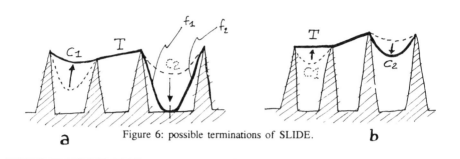

Figure 6: possible terminations of SLIDE.

this phase, the preconditions and quantity conditions of SLIDE have been recognized true, leading to the deduction that the string is sliding at a certain speed (RELATIONS field) and the length of C1 and C2 are changing (INFLUENCES field). The only way to alter this evolution, where two catenaries (IV's) are modifying their shape under the INFLUENCE of a process, is to activate a new process which adds forces that can make the quantity conditions of SLIDE become false. Therefore the intensity difference between F't and F2 must be compensate. Being both F't and F2 applied at a2, the point where a robot might grasp C2 is also individuated. Determining the exact intensity of the force to apply is of course a problem which goes beyond this kind of analysis.

Summary

The qualitative modelling of flexible objects deserves certainly more attention than it has got until now. We have presented a description which exploits the process – based ontology suggested by Forbus with the QPT. The advantages of this approach in manipulating nonrigid objects and its elements have been presented and discussed. It has been shown how qualitative reasoning about the behavior of a string as a whole and detailed analysis of string segments can be made possible. A relevant simplification regarding the domain dynamics is the hypothesis of aristotelian motion, which is suggested by the observation of the behavior of real strings. Further work is to be devoted to verify the completeness of the model through different experimental benchmarks.

APPENDIX

A list of all the entities defined in the model is given below.

OBJECTS
String, force, support, string_unit.

INDIVIDUAL VIEWS
Hanging, laid, catenary, vertical_cat, weighting_cat, taut, constraint.

ENCAPSULATED HISTORIES
hanging_to_vertical, hang_weight, catenary_to_vartical, catenary_to_taut, impact, caten_double, sustaining, taut_to_catenary.

RELATIONS
ang, appl, ascendent, direction, descendent, distance, endpoints, inclination, latitudinal, linked, median, tilted_right, tilted_left, orientation, horizontal, physic_cont, projection, verse, vertical.

PROCESSES
Slide, motion, support_movement.

1. P. Hayes, *Naive Physics 1: Ontology for Liquids,* Univ. de Geneve (1978). Memo 35, Institut pour les Etudes Semantiques et Cognitives

2. J. deKleer and J. S. Brown, ``Assumptions and Ambiguities in Mechanistic Mental Models,'' in *Mental Models,* ed. A. Stevens and D. Gentner,Erlsbaum (1983).

3. K. Forbus, ``Qualitative Process Theory,'' *Artificial Intelligence* 24(1984). Special Issue on Qualitative Reasoning

4. L. Gambardella, F. Gardin, and B. Meltzer, *Analogical Representation in Reasoning Systems,* to appear ().

5. K. Forbus, ``The Role of Qualitative Dynamics in Naive Physics,'' in *Formal Theories of the Commonsense World,* Ablex Publishing Co. (1985).

THE ALTERNATIVES ALLOWED BY A RECTANGULARITY POSTULATE,
AND A PRAGMATIC APPROACH TO INTERPRETING MOTION.

R.Cowie

Department of Psychology,
Queen's University, Belfast, Northern Ireland.

Perkins (1983) has commented that machine vision research
tends to produce a "physicist's style" which is quite unlike the
"pragmatic style" of human vision. It is easy to sympathise with this
kind of intuition, but developing it in any depth presents problems.
Not the least is that few systems offer a clear picture of what might
be involved in a "pragmatic style". This paper reports work aimed at
illustrating one possible form of "pragmatic style".

The area to be studied is the interpretation of visual motion.
Research in this area epitomises the "physicist's style" with its
elegant exploitation of the rigidity postulate either on its own
(Ullman, 1979; Longuet-Higgins and Pradzny, 1980), or in conjunction
with comparably abstract assumptions about the form which motion
takes (Webb and Aggarwal, 1981; Hoffman and Flinchbaugh, 1982). But
for human observers rigidity seems to be only one factor in the
interpretation of moving displays, even when the rigidity postulate
would in principle determine interpretation. A considerable range of
factors interact with it. Rigid interpretation can be promoted by
initial expectations consistent with it (Wallach and O'Connell,
1953). Conversely it can be blocked by weakly based initial
presumptions which conflict with it, as in Mach's example of the
reversed folded card. Lighting and occlusion cues can lead us to see
non-rigid interpretations where rigid ones are possible (Sperling,
Pavel, Cohen, Landy and Schwartz, 1983). Bradley (1983) found that
subjects watching simple moving structures judged lengths and angles
in them better when they controlled, and therefore had independent
knowledge of, the objects' motions. Regularity of form improved
subjects' judgements too. Form is also implicated in the "Rubber
Rhomboid" effect (Cowie, forthcoming), where a non-rectangular
parallelopiped is rotated slowly and appears to deform as it moves.
Green (1961) also found form a factor in promoting rigid
interpretation, as were regularity of motion and the presence of
visible connections between the points in the structure.

Ullman (1979) has handled this kind of evidence by attributing it to
a "motion from structure" analysis, in which an object's form is
inferred from a single view and conclusions about its motion follow
passively. But logically, this is not the only alternative to his
rigidity-based "structure from motion" scheme. The intermediate
possibility is that interpretation reflects assumptions about both

form and motion. Assumptions about form could constrain interpretation without determining it, leaving considerations involving motion to take up the slack: or assumptions about form could determine interpretation only so long as they led to acceptable conclusions about motion. There is good evidence that at least one of these happens in the best documented anomaly of motion perception, the set of illusions produced by rotating trapezia. These will be given special consideration in this paper.

There are three main indications that a pure structure from motion scheme is not the source of our problems with rotating trapezia. Two of these hinge on the relationship between apparent reversals and "static slant", the slant which observers impute to a trapezoidal object on the basis of a single, static view. In a stucture from motion scheme, this relationship would be simple and direct. However Braunstein and Stern (1980) showed that some factors affect judgements of slant from a single view, but not apparent reversals. In addition Mitchell and Power (1983) showed that apparent reversals occur even with "paradoxical" objects which on the basis of static slants should appear to rotate through 360 degrees rather than oscillating. The same authors also showed that apparent reversal frequency correlates highly with a kinetic variable involving the (predicted) apparent speed of a trapezium approaching a potential reversal point. Hence it seems fairly clear that the analysis which we apply to trapezia involves what may be called a compound scheme, involving expectations linked to both form and motion.

One might assume that such schemes were at most a peripheral stop-gap, used only when impoverished stimuli preclude more elegant approaches. But the evidence suggests otherwise. Rotating trapezia continue to cause us problems when they are modified so that various kinds of rational, uniform procedure would produce correct interpretations. Illusory reversals persist when viewing is binocular (Ames, 1951). With due respect to Ullman (1979), adding texture (in his sense) does not reduce reversals. Zegers (1964), whom he cites, showed only that texture did not significantly increase reversals, and in some figures oscillation certainly can be induced by texture designed to give false impressions of static slant (Borjesson, 1971; Power and Day, 1973). Adding a bar to make the moving structure three-dimensional rather than laminar does not eliminate the problem. Reversals may persist (Ames 1951), or the bar may appear to speed and slow relative to the trapezium, so that the (objectively) rigid configuration is perceived as non-rigid (Mitchell, 1985). On the other hand, reversals are effectively eliminated when the axis of rotation is tilted (Zegers, 1964: Epstein, Jansson, and Johansson 1968) or the object's subtense is increased (Zegers, 1964). These effects do not depend on the existence of texture in the object, which refutes Ullman's explanation of them, so that the visual system's successes in this area appear as puzzling as its failures.

In total, such evidence suggests that human vision may incorporate a compound scheme for interpreting motion which it uses quite extensively, and which can take into account a considerable range of prompts and hints relevant to determining form, or motion, or both.

This makes it distinctly interesting that the kernel of such a scheme falls naturally out of a priori reasoning about deploying the rectangularity assumption, which is the assumption about form most consistently implicated in motion perception. That is the finding described in the central part of this paper. The discussion falls into two parts. The first considers the interpretation of a single changing angle which is presumed to represent arms at right angles to each other. The second considers much more sketchily how the results of this analysis could be used in the interpretation of a display containing many angles.

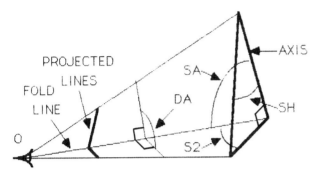

Figure 1. Terms used to describe the interpretation of a projected angle representing arms which meet at right angles.

The approach rests on discovering the possibilities which follow when projected lines are assumed to represent a right angle, and developing representations which make these apparent. (It will be assumed that projection is perspective, though parallel projection could be handled as a special case.) Figure 1 illustrates the basic terms involved in the analysis. One of the object's arms is designated the axis. The (presumed) angle between the axis and the line from viewer to (presumed) right angle will be called SA (for slope of the axis). The other main attributes of the structure follow easily from a value of SA. Calculation is simplest if instead of using the projected angle directly, one uses the angle between the two planes each of which contains one arm of the angle and the viewer. This will be called DA, the dihedral angle. The intersection of these planes - which passes through the viewer and the (presumed) right angle - will be called the fold line. Now the angle S2 between the second arm and the fold line is given by

$$\cos DA = - \cot SA \cdot \cot S2.$$

Given SA and S2 it is trivial to complete description of the structure comprising the viewer and the (presumed) right angle up to a scale factor, which may be set by letting the length of the axis arm equal 1. A particularly useful property is the tangent of the angle SH between the axis and the line joining the outermost points of the two arms at right angles. This is called SH (for "shape") because it specifies the shape of the triangle defined by the two arms flanking the (presumed) right angle.

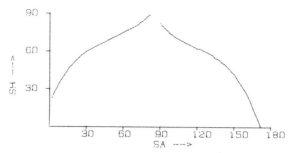

<u>Figure 2.</u> Profile expressing the possible interpretations of given lines assumed to represent a right angle.

A single view has a range of interpretations associated with it, each stemming from a different hypothesis about SA. This range can be summarised by a profile which shows the value of SH associated with each possible value of SA. Figure 2 illustrates. There are two complications associated with profiles like this, both easily handled. First, some values of SA have no associated values of SH because they imply an arm which would intersect the line from the viewer to the end point of the arm on the wrong side of the viewer, as illustrated in figure 3. These can be left as blanks in the profile. Second, if DA is 90 degrees then the only constraint is that either SA or S2 must be 90. Hence the profile representation breaks down when DA is 90 and SA is hypothesised to be 90. In that case S2 can have a range of values. The case where DA is 90 will be ignored in the immediately following discussion, but it has a special significance which will be considered later in the paper.

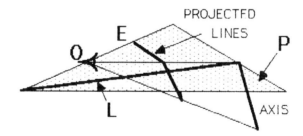

<u>Figure 3.</u> An untenable hypothesis about axis orientation. The second arm must lie in the shaded plane P. L is the line at right angles to the (conjectured) axis which lies in P, but it cuts the line OE on the wrong side of the observer O.

The interpretations associated with a sequence of views can be summarised by stacking together the profiles associated with these views. This defines an undulating and cratered surface. The most convenient way of depicting it is illustrated in figure 4. Here time is represented along the vertical axis and SA along the horizontal. The number at each point is the associated value of SH divided by 10 and rounded down. This display can be called a right angle spectrum.

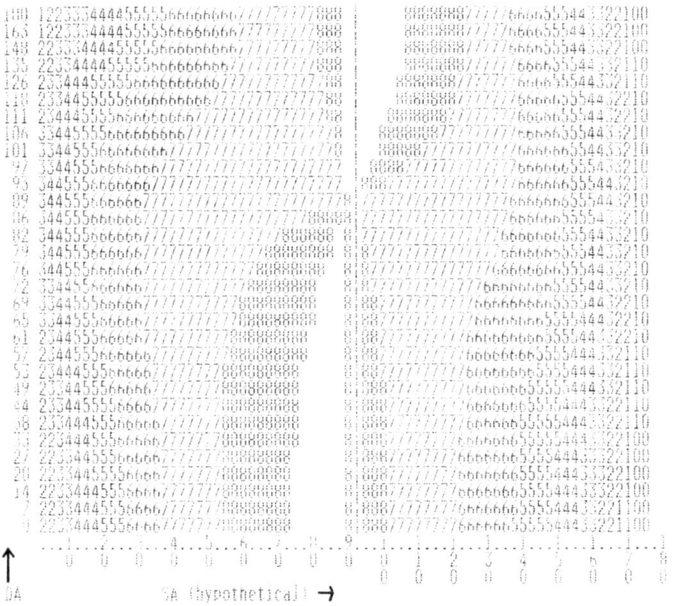

Figure 4. Right angle spectrum produced by an arm rotating about an axis at right angles to it. Here SA=73°, SH=120°, and right angle distance = 7.2 (letting axis length = 1).

A right angle spectrum, rather like the gradient space, is a device for making possibilities explicit. It would appear to be well suited to biological machinery, with its highly parallel structure. It also lends itself to various methods of settling on a particular interpretation. These methods involve selecting a path down the spectrum, from top to bottom.

The familiar rigidity constraint involves keeping SH constant, i.e. tracing a path which is always at the same height on the three-dimensional surface. Another natural constraint is that motion should be confined to one arm of the right angle, i.e. the axis should remain steady. This involves keeping paths through the spectrum as nearly vertical as possible. It is apparent that although neither constraint on its own determines a unique path, the two together usually do. A path which satisfies both constraints fully is of course distinctly interesting. In principle the column at SA=120° in figure 3 should do this, but the cumulation of small errors in calculation makes SA seem to vary along this course - which suggests why a flexible approach to path choice is advisable.

This general approach can be extended by using properties which are not declared in the spectrum as defined above, but are linked to it. Each point in the spectrum has associated with it a distance from viewer to right angle, and an angle between the plane containing the right angle and the plane from viewer to axis. These may be called RD, the right angle distance, and PR, the plane rotation angle

respectively. Figure 5 illustrates. Preferring to keep right angle
distance constant is a component of preferring to keep the axis
still. With plane rotation angle, the natural preference is that it
should change steadily. This preference may well be important in
human vision, since Mitchell and Power argue convincingly that
apparent slowing of rotation precipitates apparent reversals in
rotating trapezia.

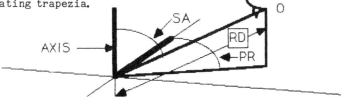

Figure 5. An object-centred frame of reference illustrating
plane rotation angle (PR), right angle distance (RD), and SA.

The comments above tacitly assume that the observer is static, and
that optical motion is due to the external structure. But the
quantities involved are equally meaningful, and sometimes more so, if
the observer is moving. A particularly satisfying inversion arises
when the axis is a vertical edge and the plane containing the angle
is taken as an external reference plane. Then SA defines the
elevation of the line of sight to the right angle (in fact SA
complements the angle of elevation), and the rotation angle defines
angle of regard projected on the horizontal plane. Under some
circumstances these may amount to object-centred co-ordinates for the
viewer in a rather natural form, as figure 5 suggests. The main point
here is that knowledge about the observer's movements, even in a
relatively crude form, could feed into the selection of constraints
to be used in selecting a path through a right angle spectrum. This
would also apply when the observer was controlling an object's
motion, as in Bradley's experiment.

A different type of constraint emerges where three lines meet at a
point, and are presumed to represent two arms at right angles to a
third (which will be designated the axis). The easiest constraint to
use here is that at any given time, descriptions of the two
(presumed) right angles must agree on the orientation of the axis.
This constraint can be exploited by extending the information
associated with each individual angle. At any given instant, one can
plot above each value of SA the value or values of SA in the previous
sample which are associated with the same value of SH. This gives
rise to a graph specifying pairs of values of SA in consecutive views
which are associated with rigid interpretations. Figures 6(a) and (b)
illustrate. Although the graphs theoretically consist of lines, in
practice there are ribbons associated with negligible change in SH
(The criterion for plotting a point in figure 6 is that values of SH
should agree within 1°.) Not surprisingly, the ribbons tend to
approximate two diagonals, one (from bottom left to top right)
indicating that axis orientation has remained relatively stable and
the other indicating that its second position is roughly a mirror
image of its first. The hashed bands mark gaps in the relevant
profiles, where no meaningful interpretations exist.

Superimposing two such graphs, as in 6c, expresses the fact that if two (presumed) right angles share an axis, then descriptions of them must agree on the orientation of the axis. This can be regarded as an application of the expectation that viewing position will be general (Cowie, 1983). So wherever the plotted ribbons overlap, there is an interpretation which involves two arms at right angles to a third, and in which all three remain the same length. This is not necessarily a fully rigid interpretation: nothing constrains the first two arms to maintain the same angle to each other.

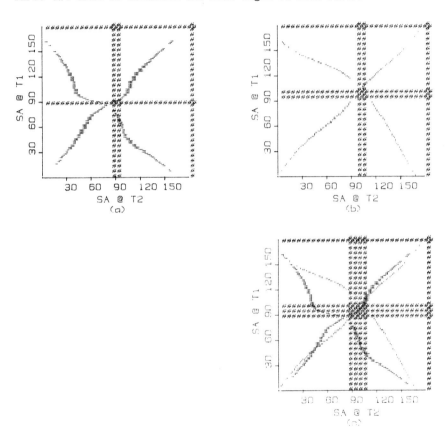

Figure 6. Rigid interpretations associated with consecutive views of a triangle. A moving triangle observed twice defines alternatives of the sort shown in (a) and (b). Panel (c) deals with two views each showing two triangles which share an edge. Common interpretations are found by superimposing the graphs associated with the individual triangles. The orientations for the shared edge are points where the ribbons intersect. Here the triangles shared an axis at SA=120°, and rotated through 20°. One (shown in (b)) started at the sagittal, the other started 80° further on. Other parameters were as in figure 4.

116   R.Cowie

This treatment of paired angles retreats from the strategy of
allowing multiple constraints to affect interpretation, and moves
back towards the more familiar strategy of letting a few constraints
determine interpretation. However that can be avoided. Figure 7
illustrates an open-ended representation analogous to those of figure
6. It is created by plotting above each possible value of SA at time
2 how much shape change each possible value of SA at time 1 would
imply, instead of making a mark only if shape change is zero as
happened in figure 6. Superimposing arrays like this (by adding the
numbers at each point) specifies how much shape change is associated
with each possible hypothesis about consecutive orientations of an
axis which two arms (ex hypothesi) meet at right angles. The same
kinds of criteria as were considered before can be invoked in
choosing a path through a sequence of such arrays.

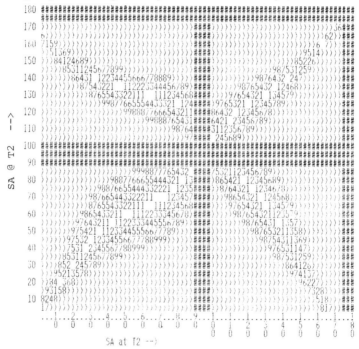

Figure 7. Degrees of change in SH associated with possible
hypotheses about SH in two successive views. For visibility
change <1° is marked with a blank, change >9° with a carat.
The triangle involved was like those of figure 6, but it
started at 20° from the sagittal.

This discussion does not exhaust the problem of interpreting a single
moving vertex which is presumed to contain right angles, but it
introduces the main ideas. Attention can now be turned to how these
ideas relate to interpreting displays containing multiple vertices.
This raises two main issues. These have not been explored rigorously,
but some key points about them are reasonably clear.

The first issue is how to identify right angles. At first sight it seems attractive to imagine a network in which all angles are initially assumed to be right, with an agreed constellation emerging as competing claims inhibit each other and compatible claims reinforce each other. The minimum compatibility constraint between local interpretations is that they should agree on the orientation of any edge which they share. It would be natural to prefer constellations which allowed preservation of lengths and regularity of movement. One might also prefer edges to maintain their relative orientations. A special form of this preference, that edges should be coplanar or in parallel planes, is particularly accessible because it can be expressed in terms of the plane rotation angle. But because of the degrees of freedom associated with each vertex individually, it is difficult to see how such a network could be set up.

Considering this difficulty suggests another possibility. This is that postulates about right angles, and other related structural regularities such as parallelism, would not initially be indiscriminate. Instead selective and extended structural hypotheses would be imported, and right angle spectra would (at least for the most part) be involved in assessing these globally. This has some immediate psychological attractions. Our interpretation of static pictures indicates that we do have the ability to develop such hypotheses; evidence such as Mach's card suggests that they can dominate our interpretation of motion; and intuitively it sometimes does seem that the ability to detect problems with an existing interpretation is indeed separate from the ability to find a better solution - in contrast to the pattern in schemes like Waltz's (1975), where solutions emerge smoothly from a process of elimination.

The second general issue is how to propagate the implications of a decision that a particular scene fragment represents right angles moving in particular ways. Two types of propagation are of interest.

The first type may be called structure independent. Decisions about a moving right angle are potentially a key to wider analysis based on the rigidity postulate. This is because determining the position of a right angle in successive views determines the course which the observer must have followed between one view and the next. More precisely, it locates both viewpoints in a single frame of reference set by the right angle. As was noted above, parameters associated with the right angle spectrum are effectively object-centred co-ordinates for the viewer's position. A viewer could use these co-ordinates to reconstruct his last position in his present frame of reference. The line through that point and his present viewpoint is therefore his net course over the time interval being considered. Once the course has been reconstructed, it is straightforward to test which points could be part of the same rigid configuration as the right angle (both images must lie in a plane which passes through the course) and to reconstruct their distances from the present viewpoint (using the scale factor associated with the right angle spectrum).

The second type of propagation is structure dependent. If, as suggested above, the right angle spectrum supplements schemes concerned with picking out structural regularities, then it may often be unnecessary to use the structure-independent analysis described above. Particularly where not only right angles, but parallels are presumed present, slopes and lengths in a right angle which has been analysed fully may relate very simply to slopes and lengths in one which has not - equalities, addition and subtraction may be all that is needed to reconstruct the important characteristics of an edge.

These ideas seem to offer a prospect of capturing a range of characteristically human visual behaviours. Various general parallels have been noted in the discussion. But in the particular case of rotating trapezia, they promise to account quite closely for the main features of the evidence.Two main points are involved.

The first concerns the fact that at least in some sense, we appear to process the motions of rotating trapezia as if the objects were rectangular. The problem here is that there is apparently ample information to tell us that they are not rectangular, and in fact we do not think that they are. People neither think that the corner angles are right, nor do they see the massive changes in the lengths of sides that would follow from the assumption that the objects were rectangular.

Appealing to the right angle spectrum suggests the following solution. Assume that a basic path is traced through the spectrum for one angle A. Presumably the choice of path is constrained by consideration of at least one other angle. The obvious candidate is the one immediately above or below A, and the obvious form of constraint is to require that the plane rotation angle be the same in both cases - i.e. the object is laminar. This will produce a path involving large changes in SA - representing the length changes noted above - as the object rotates. The suggestion is that these changes are taken as evidence that the object is not quite rectangular, and the length changes are discounted. Conclusions about the object's orientation are retained, though.

The details of this suggestion are clearly ad hoc, but equally clearly the general approach does make sense within the general framework developed above. It embodies a general principle which was suggested above on other grounds, that detecting problems and finding genuinely more satisfactory solutions are separate issues; and it involves the kind of compromise that runs through the whole use of the right angle spectrum. And if this solution is only partly principled, no defensible alternative is even partly principled.

The second main point rests completely on basic principles. It concerns the fact that apparent reversals occur, and that various factors reduce them. An explanation of this evidence flows directly from the topology of the right angle spectrum.

The right angle spectrum of a continuously changing angle always has a blank running alongside the line SA=90. This corresponds to values of SA which imply that the second arm of the right angle is end-on to the viewer (at SA=90) or terminates behind the viewer (when SA is

just below 90 if the projected angle is acute, just above 90 if the
projected angle is obtuse). A physically meaningful path through the
spectrum cannot cross these blanks, so they split possible
interpretations into two blocks. This may explain why cues such as
interposition and relative brightness, which determine the order of
points on a line, can influence interpretation so strongly. In terms
of the right angle spectrum, their effect is to determine which side
of the central blank interpretations occupy.

In one critical case, though, the central blank vanishes. This occurs
when the dihedral angle DA=90. As DA approaches and passes through
90, the central blank shrinks to zero and then emerges on the
opposite side of the line SA=90. At this juncture, either arm can
pass through the plane at right angles to the fold line (which runs
from the observer to the point of the right angle). It is convenient,
if slightly loose, to call this the frontoparallel plane. If the path
continues on the same side of DA=90, then the axis remains on the
same side of this plane and the second arm swings through it. If the
path crosses SA=90 at DA=90, then the axis swings through the
frontoparallel and the second arm remains on the same side of it.

The key converse is that if DA never reaches 90, then both arms must
remain indefinitely locked on one side of the frontoparallel (the
same side if DA is always obtuse, opposite sides if DA is always
acute). So an angle sequence where DA never reaches 90 cannot
represent one arm rotating about the other. Oscillation is the only
possible interpretation which involves cyclic motion of one arm about
the other (presuming still that the arms are at right angles).

This accounts directly for the main outlines of the evidence on
rotating trapezia. In standard presentation conditions, DA does not
reach 90 and apparent oscillation occurs. Rotation re-emerges when
the object is made more nearly rectangular, or its vertical subtense
is increased by enlarging it or bringing it nearer, or when the axis
of rotation is tilted. The effect of all these manipulations is to
make DA pass through 90, or approach it more closely (see figure 8).

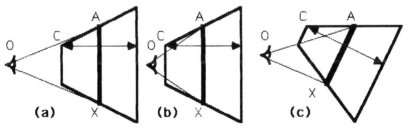

(a)   (b)   (c)

Figure 8. Rotating trapezia, depicted as they swing through
the plane OAX which contains observer (O) and axis (AX). The
arrowed line shows the path traced by corner C as it rotates.
The projection dihedral of angle CAX reaches 90° when this
path cuts the plane through OA and orthogonal to OAX. This
does not happen in standard viewing conditions (panel (a));
but it does in conditions which reduce reversals - e.g. with
reduced viewing distance (see (b)) or tilted axes (see (c)).

There is fine grain in the evidence which remains to be accommodated. A prime example is Mitchell and Power's demonstration that apparent reversal frequency relates to predicted change in apparent speed at a critical point. The accommodation seems possible in principle. The key predictions about apparent speed can be made by way of right angle spectra as well as by way of their favoured variable, and failure to reverse has an interpretation in the present context which is notably absent in Mitchell and Power's account - a path may leap the central blank around its narrowest point if this markedly improves conservation of momentum. But appealing as this kind of verbal formula may sound, translating it into a working model is clearly a long way off.

This sketch is interesting partly because it diverges from stock psychological analyses of the difference between machine and human vision. According to Perkins, human vision's "pragmatic style" involves exploiting strong, rather improbable regularities rather than using analyses of any generality. This view is echoed in the major psychological accounts of rotating trapezia (Braunstein, 1976; Mitchell and Power, 1983). They propose that vision errs by abusing heuristics which are accurate only when applied to a rectangle mounted on an axis which is vertical and parallel to one side of the rectangle, and when motion between observer and object consists purely of the latter rotating about its axis at a steady speed. These requirements are common to both accounts: each also incorporates requirements specific to itself.

Human vision may consist of components like this, but it is difficult to see how they could be welded into such a reliable general-purpose system. The research reported here stemmed from that difficulty, and set out to find whether the evidence was consistent with a more general type of mechanism. The demonstration that it is in the case of rotating trapezia raises doubts about Perkins' more general case. Again, Perkins proposes that declaring alternative possibilities, as in Kanade's use of gradient space, is a strategy avoided by human vision. Yet here the right angle spectrum, which does exactly that, offers a key to a range of apparently human-like behaviours.

The proper conclusion is not, of course, that the kind of analysis developed here is the true mirror of human vision. A great deal obviously needs to be done before it is even clear whether it is viable. However it does appear that before we can come to well-founded conclusions about the style of human vision, we need to understand the logical possibilities a great deal better.

Acknowledgements are due to Rex Mitchell for discussions which provoked this approach.

REFERENCES

Ames,A. (1951). Visual perception and the rotating trapezoid window. Psychological Monographs 65, whole no. 324.

Borjesson, E. (1971). Properties of changing patterns evoking visually perceived oscillation. Perception and Psychophysics 9, 303-308.
Bradley,D. (1983) The role of action in perception. Irish Psychologist 9(8), 35-36.
Braunstein,M.L. (1976). Depth Perception through Motion. London: Academic Press.
Braunstein,M.L. and Stern,K.R. (1980). Static and dynamic factors in the perception of rotary motion. Perception and Psychophysics 27, 313-320.
Cowie,R. (1983). The viewer's place in theories of vision. Proceedings of the 8th I.J.C.A.I. Karlsruhe: I.J.C.A.I.. pp. 952-958.
Cowie,R. (forthcoming). Rigidity and the Rubber Rhomboid. To appear in Perception and Psychophysics.
Epstein,W., Jansson,G., and Johansson,G. (1968). Perceived angle of oscillatory motion. Perception and Psychophysics 3, 12-16.
Green,B.F. (1961). Figural coherence in the kinetic depth effect. Journal of Experimental Psychology 62, 272-282.
Kanade,T. (1981). Recovery of the three-dimensional shape of an object from a single view. Artificial Intelligence 17, 409-460.
Hoffman,D.D. and Finchbaugh,B.E. (1982). The interpretation of biological motion. Biological Cybernetics 42, 195-204.
Longuet-Higgins,C.M. and Pradzny, K. (1980). The interpretation of a moving retinal image. Proc. Royal Soc. Lond. B 208, 385-397.
Mitchell,R. (1985). The rotational transformation with rectangles: invariance with shape and distance. Bulletin of the British Psychological Society 38, A96.
Mitchell,R. and Power,R. (1983). Apparent reversal frequencies in squares and trapezia: the effect of variant and invariant information. Perception 12, 375-392.
Perkins, D.N. (1983). Why the human perceiver is a bad machine. In J.Beck, B.Hope, and A.Rosenfeld (eds.) Human and Machine Vision. New York: Academic Press. pp. 341-364.
Power,R.P. and Day,R.H. (1973) Constancy and illusion of apparent direction of rotary motion in depth: tests of a theory. Perception and Psychophysics 13, 217-223.
Sperling,G., Pavel,M., Cohen,Y., Landy,M.S., and Schwartz,B.J. (1983). Image processing in perception and cognition. In O.Braddick and A.Sleigh (eds.) Physical and Biological Processing of Images. Berlin: Springer.
Ullman,S. (1979). The Interpretation of Visual Motion. Cambridge, Mass.: M.I.T. Press.
Wallach,H. and O'Connell,D.N. (1953). The kinetic depth effect. Journal of Experimental Psychology 45, 205-217.
Waltz,D.L. (1975). Interpreting pictures of scenes with shadows. In P.H. Winston (ed.) The Psychology of Computer Vision. New York: McGraw Hill.
Webb,J.A. and Aggarwal,J.K. (1981). Visually interpreting the motion of objects in space. Computer 14(8), 40-46.
Zegers,R.T. (1964). The reversal illusion of the Ames trapezoid. Transactions of the New York Academy of Science 26, 377-400.

Reasoning and
Formal Systems

Semantic Tableaux as a framework for Automated Theorem-Proving

Steve Reeves

Department of Computer Science and Statistics
Queen Mary College
University of London
LONDON
E1 4NS

ABSTRACT

The aim of this paper is to give examples to show the utility of semantic tableaux as a general framework for developing theorem-provers. The paper has two parts: first, the main part presents an incorporation of unification into a semantic tableau theorem-prover together with proofs of soundness and completeness and a brief look at the results of an implementation that we have developed based on this method; we also aim to give a bibliography both of other work on semantic tableaux that has been done in the last 30 years with reference to theorem-proving, and on work which shows how semantic tableaux have been used as a framework for presenting proof systems for logics other that first-order, classical logic.

Keywords: theorem-proving, semantic tableaux, non-classical logics, unification.

1. Introduction

Semantic tableaux have been used as a uniform method for producing proofs in logic and meta-logic extensively in the fields of mathematical and philosophical logic. However, though there have been several (rather isolated) workers in automatic theorem-proving who have used semantic tableaux and found them a natural, expressive and adaptable framework (we review some of this work below), they have been little used by the automated reasoning community at large.

We first give a very brief view of the standard semantic tableau method. This has been presented many times before ([2],[12],[18],[19],[22] for example) and so is already very accessible - hence the brief coverage. We then introduce the less discussed use of unification to cut down the size of the search space and present proofs of soundness and completeness of the method for this case, together with an algorithm which forms the basis of our implementation.

Following this we give a brief bibliography of the work of many people which give examples of how ideas from resolution-based theorem-proving have already been expressed in a semantic tableau framework and how the basic method can be made more efficient (with or without completeness), and of how semantic tableaux have been used to present modal and intuitionistic logics.

In what follows we refer the reader to [19] or [22] for proofs that we omit, since these will all be well-known and accessible results.

2. The Origins of Semantic Tableaux

Semantic tableaux were introduced and discussed (under that name) by Beth [2] in 1955. In the same year a paper by Hintikka [11] introduced model sets, and we can now see that these two ideas are essentially the same.

A unified treatment of their work is given in Smullyan [22] (whose presentation we shall follow) which also generalizes the approach to provide a uniform method for describing other first-order logics.

In Hintikka [11] the idea of tableaux is motivated by trying to resolve problems in the foundations of mathematics. There were many arguments taking place about the form and content dualism

125

in mathematics. Philosophers of mathematics were arguing about whether valid mathematics is that in which, though the meaning of a proof may not be clear, certain formal rules which are known to be sound are followed to produce the proof; or whether one should only allow arguments whose meaning were quite, intuitively, clear to everyone, i.e. should it be the form or should it be the meaning of an argument which is obviously valid ?

The two main schools which attempted to bridge this gap were the intuitionists of Brouwer and the finitists of Hilbert.

Brouwer held that any concept that was not intuitively justifiable could not be accepted. He thus proposed, for instance, that the principle of the excluded middle, that any property is either true or false (independently of whether we can tell which pertains), was not valid. His mathematics, based on these principles, was not very satisfactory to the classical mathematicians in that many of the 'intuitively invalid', though frequently used, tools of most mathematicians would not be available. However, more recent work, notably that of Bishop [6] has 'recovered' much of what was 'lost' at least for a constructive mathematician, if not for a full-blown intuitionist. This work has recently become very important in Computer Science as embodied in the work of Per Martin-Löf [14].

Hilbert, on the other hand, agreed that some mathematical statements are intuitively valid and that the rest of mathematics is only valid if a finite argument, i.e. using a finite set of fixed rules, can lead to a derivation which proceeds from the agreed set of truths to the required set. Hilbert's scheme failed because it can be shown that without using non-finitist arguments the consistency of the scheme cannot be proved. Thus, neither system was successful.

This being the position in the philosophy of mathematics, Beth and Hintikka discussed the problem by restricting attention to a well-known and non-trivial part of mathematical reasoning, predicate calculus. As Hintikka says [11],

"...there are certain simple ways of reasoning, the meaning of which seems clear to us; there are other methods of argument whose interpretability is not obvious. What the intuitionists and formalists tried to do is to reduce the later to the former. Instead, one should try rather to understand both kinds of proof".

Thus, in each paper the respective author brings a constructive method to bear on what was, at the time, a problem usually tackled by a formal derivation approach. That is, proofs in predicate calculus were conducted by taking the initial axioms and trying to derive the sentence to be proved by applying a fixed set of rules. That is not to say that the approach of constructing sets of individuals within which sentence in question was either true or false was unknown. Rather, it was thought to be highly dubious since to treat quantification a set of individuals has to be postulated and for many sentences this set has to be infinite. Thus the set-theoretic treatment insisted on the infinite sets being on an equal footing with the finite sets in that they are whole, finished entities.

However, as Hintikka points out, the purely formal system is also open to question. Each axiom-rule system has to be shown to be consistent to be of any use and these various proofs of consistency had to appeal to set-theoretic notions of validity.

Thus, both the authors cited here introduced strictly constructive notions, to satisfy the arguments against holding finished infinite sets as reasonable objects, which can be shown to allow derivations as in the then more usual formal systems.

They use a step-by-step construction of the sets of individuals within which the sets are to be interpreted so that as many individuals can be presented as may be required for a particular proof, but the concept of a finished infinite object can be avoided. As an aside, this idea is rather suggestive of the way computable functions are treated nowadays. Although a function, in many cases, can only be represented by a infinite set of argument-value tuples we think of a computation gradually showing us more and more of the function without us having to think in terms of infinite objects. We usually say that they are potentially infinite.

As I hope will be shown, these philosophical points, which need not concern anyone who is willing to take the soundness of the methods on trust, lead to a very natural and simple method for reasoning in predicate calculus.

3. Notation and Definitions

We assume the usual definition of a *first-order language* and of *terms (from a set* T*), names (from a set* N*), formulae, sentences, atoms* and *literals*.

Definition 3.1 an *interpretation of a (first-order) language L* over a universe U is a structure $I_L^U = <\phi^U, \psi^U>_L$ where

U is a non-empty set of individuals
$\psi^U : P_n \to U^n$
$\phi^U : N \to U$
$\phi^U : F_n \to (U^n \to U)$

Here, P_i is a set of predicate symbols of arity i, F_i is a set of function symbols of arity i and we usually drop the superscript U when no ambiguity can arise.

We now have a basis upon which to judge the truth or falsity of a sentence in a particular language. We have to extend ϕ so that $\phi : T \to U$, i.e. when applied to a term in T, ϕ produces as its value an object from U. So, $\phi(g(t_1,...,t_m)) = \phi(g)(\phi(t_1),...,\phi(t_m))$ for each $g \in F_m$, $t_i \in T$. We call ϕ a *valuation*.

To make any valuation complete each variable must also denote an object in U. It is the assignment of objects to variables that is the central problem in providing valuations for sentences. We can now give a hierarchy by which the truth of a sentence in L can be determined.

Definition 3.2 a valuation ϕ *satisfies* a sentence in an interpretation according to the following:

1) for any predicate $Q \in P$, $\alpha(Q)=n$, ϕ satisfies $Q(t_1,...,t_n)$, $t_i \in T$, iff $\psi(Q)(\phi(t_1),...,t_n))$ is true in U.

2) ϕ satisfies $\neg S$ iff ϕ does not satisfy S

3) ϕ satisfies $(S \wedge T)$ iff ϕ satisfies S and T,
    ϕ satisfies $(S \vee T)$ iff ϕ satisfies S or T, or both,
    ϕ satisfies $(S \to T)$ iff ϕ satisfies $(\neg S \vee T)$,
    ϕ satisfies $(S \leftrightarrow T)$ iff ϕ satisfies $(S \wedge T) \vee (\neg S \wedge \neg T)$

4) ϕ satisfies $(\exists x)S$ iff there is an $x \in V$ such that ϕ satisfies S when x is replaced in S by the denotation of $\phi(x)$,
    ϕ satisfies $(\forall x)S$ iff ϕ satisfies $\neg(\exists x)\neg S$

Now, a sentence is *true* in an interpretation I iff every valuation in I satisfies it and finally, a sentence is *valid* iff it is true in every interpretation of its language L. Conversely, a sentence is *false* in an interpretation I iff no valuation in I satisfies it and it is *unsatisfiable* iff it is false in every interpretation. Some sentences, of course, may not be unsatisfiable. It will be convenient to call such sentences *satisfiable*.

Definition 3.3 a *model* is an interpretation of a sentence which makes the sentence true. We also extend this to sets of sentences, Σ, and say that M is a model of Σ iff M is a model for every sentence in Σ.

4. Semantic Tableaux for Predicate Calculus

Given a set of sentences of the form $\Sigma \cup \{\neg S\}$ we want to show that no model for it exists, i.e. no valuation satisfies it in any interpretation. We say that the semantic tableau method is a 'proof by refutation' method. We want to refute the existence of a model of $\Sigma \cup \{\neg S\}$.

We display the evolution of the refutation in the form of a binary tree and refer to *paths* in such a tree with the usual meaning.

Definitions 4.1. the sentences in $\Sigma \cup \{\neg S\}$ are the *initial sentences* of the tableau. $\Sigma \models S$ is the *entailment represented by* the initial sentences given by $\Sigma \cup \{\neg S\}$.

Definitions 4.2. a path is *closed* iff it contains a sentence and its negation. A tableau is *closed* iff all its paths are closed.

Definition 4.3

1) a sentence S of the form $A \wedge B$, $\neg(A \vee B)$ or $\neg(A \to B)$ will be called a *sentence of type* α. If S is of the form $A \wedge B$ then α_1 denotes A and α_2 denotes B. If S is of the form $\neg(A \vee B)$ then α_1 denotes $\neg A$ and α_2 denotes $\neg B$. If S is of the form $\neg(A \to B)$ then α_1 denotes A and α_2 denotes $\neg B$.

2) A sentence S of the form $A \vee B$, $\neg(A \wedge B)$, $A \leftrightarrow B$, $A \to B$ or $\neg(A \leftrightarrow B)$ will be called a *sentence of type* β. If S is of the form $A \vee B$ then β_1 denotes A and β_2 denotes B. If S is of the form $\neg(A \wedge B)$ then β_1 denotes $\neg A$ and β_2 denotes $\neg B$. If S is of the form $A \to B$ then β_1 denotes $\neg A$ and β_2 denotes B. If S is of the form $A \leftrightarrow B$ then β_1 denotes $A \wedge B$ and β_2 denotes $\neg A \wedge \neg B$. If S is of the form $\neg(A \leftrightarrow B)$ then β_1 denotes $\neg A \wedge B$ and β_2 denotes $A \wedge \neg B$.

3) A sentence of the form $(\forall x)F$ will be called a *sentence of type* γ. For such a sentence $\gamma(a)$ denotes $F\{<a,x>\}$, i.e. the sentence F with x uniformly replaced by a.

4) A sentence of the form $(\exists x)F$ will be called a *sentence of type* δ. For such a sentence $\delta(a)$ denotes $F\{<x,a>\}$.

Lemma 4.4 For any interpretation I_U the following hold, by the usual definition of the connectives :-

C_1: a sentence of the form α is true iff both α_1 and α_2 are true.

C_2: a sentence of the form β is true iff β_1 or β_2 is true.

C_3: γ is true iff $\gamma(a)$ is true for every a such that $\phi(a) \in U$

C_4: δ is true iff $\delta(a)$ is true for some a such that $\phi(a) \in U$

Definition 4.5 A tableau T_2 is a *direct extension* of a tableau T_1 if it can be obtained from T_1 by application of one of the following rules, where p_l is a path in T_1:

D_1) if some α occurs on path p_l with leaf l, then join α_1 to l and α_2 to α_1 to get T_2.

D_2) if some β occurs on path p_l with leaf l then add β_1 as the left successor of l and β_2 as the right successor of l to get T_2.

D_3) if some γ occurs on a path p_l with leaf l, then join $\gamma(a)$ to l, for some name a that occurs in a sentence on p_l

D_4) if some δ appears on a path p_l with leaf l, then join $\delta(a)$ to l, for some name a which does not occur in any sentence on p_l.

Lemma 4.6 For any satisfiable set S of sentences the following hold:

(i) S_1: if $\alpha \in S$ then $\{S, \alpha_1, \alpha_2\}$ is satisfiable.

S_2: if $\beta \in S$ then $\{S, \beta_1\}$ or $\{S, \beta_2\}$ is satisfiable.

S_3: if $\gamma \in S$ then $\{S, \gamma(a)\}$ is satisfiable for every name a occurring in any element of S

S_4: if $\delta \in S$ then $\{S, \delta(a)\}$ is satisfiable as long as a does not occur in any element of S

ii) If a tableau T_2 is a direct extension of a tableau T_1 which is true under $<\phi,\psi>$, then T_2 is true under $<\phi',\psi>$, where ϕ' conservatively extends ϕ.

Corollary 4.7 If the initial sentences of a tableau T generated by D_1, D_2, D_3 and D_4 are satisfied then T must be true.

Soundness Theorem 4.8 Any entailment provable by a tableau must be valid.

Definition 4.9 S is a *model set* in the following cases:

H_0: No atomic sentence and its negation are both in S

H_1: if $\alpha \in S$, then $\alpha_1, \alpha_2 \in S$

H_2: if $\beta \in S$, then $\beta_1 \in S$ or $\beta_2 \in S$ or both

H_3: if $\gamma \in S$ then $\gamma(a) \in S$ for every name a occurring in the sentences of S

H_4: if $\delta \in S$ then $\delta(a) \in S$ for some name a occurring in the sentences of S.

Lemma 4.10 If S is a model set then it is satisfiable.

Definition 4.11 A path in a tableau is *complete* iff

i) for every α on p, both α_1 and α_2 are on p

ii) for every β on p, at least one of β_1 or β_2 is on p.

iii) if for every γ on p and every name a in any sentence on p the sentence γ(a) appears on p

iv) if for every δ on p, δ(a) appears on p for some name a.

Definition 4.12 A tableau is *completed* iff every path is either closed or complete.

Definition 4.13 $\Sigma \models S$ is *provable* iff there is a tableau with initial sentences $\Sigma \cup \{\neg S\}$ which has an extension which is closed.

Definition 4.14 the *universal rule* is:

Given a sentence of the form $(\forall x)(F(x))$ with x free in $F(x)$ on an open path which contains names $n_0,...,n_{m-1} \in N$ add each of the sentences $F(x)<x,n_i>$ to the end of the path iff $F(x)<x,n_i>$ does not appear already on the path $(0 \leq i \leq m-1)$. Note for future use that the tableau has changed.

We note here that there may not be any names on the path when this rule is applied. In this case, since we are dealing with a non-empty universe, we take any name and apply the rule using this name. We shall usually take 'a' to be this first name.

Definition 4.15 the *existential rule* is:

Given a sentence of the form $(\exists x)(F(x))$ with x not bound in F on an open path which contains names $n_0,...,n_{m-1} \in N$, mark the sentence as used and add the sentence $F(x)<x,n_m>$ where $n_m \in N - \{n_0,...,n_{m-1}\}$. Note for future use that the tableau has changed.

Now consider the following algorithm:

begin
repeat
    changes ← **false**
    close each path that contains a sentence and its negation
    if all paths are closed **then deliver** "valid entailment"
    else
        if a splitting rule can be applied to a sentence **then**
            changes ← **true**
            apply the appropriate splitting rule
            mark the sentence as used
        else
            apply the existential rule
            apply the universal rule
until not changes

deliver "entailment is not valid"
end

This algorithm can be seen to achieve the required soundness and completeness conditions by noticing that at the beginning of each time around the loop the tableau is open and 'changes' is true. At the end of the loop either

i) all the paths are closed, i.e. 'success' is true (so no splitting rule has been applied) or

ii) a splitting rule has been applied (and at least one path is open) or

iii) no splitting rule is applicable (and at least one path is open) so we use either the universal rule or the existential rule.

In case (i) we are obviously finished since the tableau is closed and by the soundness theorem the entailment must be valid. In case (ii) we added a finite number of sentences to the tableau, each of which had one less connective than the split sentence that produced them.

Thus, we only re-enter the loop in cases (ii) or (iii). For case (ii), in the absence of universal sentences, the number of occurrences of connectives in unmarked sentences will be strictly decreasing so we will eventually reach a stage where no connectives appear in unmarked sentences, thus the algorithm always terminates.

We can argue for the correctness of this algorithm in the case where we do have universally quantified sentences as above though with one big difference. Because a direct extension by D_3 does not exhaust the sentence it is applied to, i.e. the sentence is not marked, we may never actually get a completed tableau in some cases. However, since any given γ will always be dealt with after some finite time (since the preceding part of the algorithm deals only with the connectives, and they are always dealt with in a finite time) it is clear that any tableau makes progress towards being completed. So, even if some path is never closed it must become nearer and nearer to forming a model set. (Smullyan [22] gives other examples of how to organise the rules so that they always lead to a model set if one is possible.) For details of our implementation of this see [19].

From this discussion we have

Completeness Theorem 4.16 If $\Sigma \models S$ is valid then it is provable.

Corollary 4.17 If $\Sigma \models S$ is valid then it is provable in a finite number of steps.

From the above it is clear that we have

Correctness Theorem 4.18 The algorithm started with initial sentences $\Sigma \cup \{\neg S\}$ will terminate with "entailment is valid" iff $\Sigma \models S$. If the algorithm terminates with "entailment is invalid" then $\Sigma \not\models S$.

5. Dummy variables

This draws on an idea first mentioned by Prawitz [17]. When one has used the method described above on many examples one begins to see an obvious proliferation of unnecessary applications of the $\forall$-rule.

Ideally what we would like to be able to do is to predict, in some way, which instantiations are necessary to obtain a proof whilst also ensuring that the method remains complete and sound. We must still have the capability to try every possible name on a given path in every universally quantified sentence on the path. This is to make sure that, following our basic rule, every sentence is allowed to express itself in every way possible in trying to build up a model. What we do, following Prawitz, is to delay any instantiations until we reach a position which allows us to predict, to a certain extent, which instantiations are required to close a path, since this is our general aim in showing $\Sigma \cup \{\neg S\}$ unsatisfiable, i.e. proving $\Sigma \models S$.

What we do is to instantiate, when the $\forall$-rule is applied, the universally quantified variable with a dummy variable. We denote these by $x_1,....$ These dummies are thus place markers which can be replaced by any name, though with the following restriction. If we have a sentence
$$(\forall x)(\exists y)(P(x,y)\wedge\neg P(x,x))$$
then the new $\forall$-rule will give
$$(\exists y)(P(x_1,y)\wedge\neg P(x_1,x_1))$$
say. Now when the $\exists$-rule is applied we get
$$P(x_1,a)\wedge\neg P(x_1,x_1).$$
We would clearly wish to substitute a for x_1 to close the path, since we would then have P(a,a) and $\neg$P(a,a), i.e. a contradiction.

However, using the original $\forall$-rule we could never have arrived at this situation because if
$$(\forall x)(\exists y)(P(x,y)\wedge\neg P(x,x))$$
had given
$$(\exists y)(P(a,y)\wedge P(a,a))$$
then this would have given, say,
$$P(a,b)\wedge\neg P(a,a)$$
but not
$$P(a,a)$$
since the $\exists$-rule requires that we instantiate with a new name. We see, therefore, that constraints must be placed on dummy variables so as to keep the rules sound. A given dummy will build up a list of constraints as it occurs in sentences to which the $\exists$-rule is applied. Thus, the example of applying the $\exists$-rule to
$$(\exists y)(P(x_1,y)\wedge\neg P(x_1,x_1))$$
will give
$$P(x_1,a)\wedge\neg P(x_1,x_1)$$
and the constraint $x_1 < a$ which says that x_1 cannot be substituted for by a at any future stage. Thus the path will not close, as required.

The normal splitting rules can then be applied to sentences containing dummies. With the original system we can see that if we have a universally quantified sentence on an open path with k names on it then we could potentially add k new sentences to the path. We would only add less than k if one or more instantiated sentences repeated a sentence that already appeared on the path. A certain number of these new sentences may be totally irrelevant when trying to close the path. With the new $\forall$-rule we get one new sentence on each open path which contains the original sentence, no matter how many names appear on that path. However, we still have the potential for the path to close in this case precisely when it would have closed with the old $\forall$-rule.

Consider the sentence A(n), say, which would have paired with $\neg$A(n) previously to close a given path, where n is a name. The old $\forall$-rule would have produced $\neg$A(n_i) for all names n_i on the path. The new $\forall$-rule will produce just $\neg$A(x_1) and by matching x_1 and n we can again close the path.

6. Extended definitions

Before going on to present a full definition of the extended method together with proofs of soundness and completeness we need to look at several new definitions. We assume that we are, as usual, working with a first-order language though we augment it now so that a dummy (variable), a member of a denumerable set of symbols $x_1, x_2,...,$ is a term. A ground term will now be a term which contains no variables or dummy variables.

Definition 6.1. an *interpretation (of a first-order language) over a universe U* is a structure $I_U =$ $<\phi_U,\psi_U>$ where U is a non-empty set of individuals, $\psi_U : P_n \to P(U^n)$ and

$$\phi_U : N \to P(U)$$
$$\phi_U : F_n \to (U^n \to U)$$
$$\phi_U : D \to P(U)$$

where P is the powerset constructor, P_i is a set of predicate symbols of arity i, N is a denumerable set of names, F_i is a set of function symbols of arity i and D is a denumerable set of dummies $x_1, x_2,....$

This is a generalisation of the usual idea of interpretation. We usually do not have any dummies and all occurrences of $P(U)$ would be U instead. We need this extra level since dummies will denote *sets* of individuals. Names will always denote singleton sets of individuals. Later we will see that certain conditions need to be put upon exactly which structures are allowable interpretations.

As usual we extend ϕ_U so that

$$\phi_U*(f(t_1,...,t_n)) = \phi_U(f)(\phi_U(t_1),...,\phi_U(t_n))$$

where for a tuple of sets $(X_1,...,X_n)$ and n-ary function F we have

$$F(X_1,...,X_n) = \{ F(x_1,...,x_n) \mid x_i \in X_i, 1 \le i \le n \}$$

as usual.

Also we have

Definition 6.2. A$(a_1,...,a_n)$ is *true* iff

$$\{(e_1,...,e_n) \mid e_1 \in \phi_U*(a_1) \wedge ... \wedge e_n \in \phi_U*(a_n) \} \subseteq \psi_U(A)$$

which we will write as

$$<\phi_U*(a_1),...,\phi_U*(a_n)> \in \psi_U(A).$$

We usually omit the subscript U and the $*$ if this causes no confusion.

We said above that we intend the denotation of a dummy to be a set of individuals. We now need to place some conditions on this to reflect the idea of constraints. If we have a dummy x with $x<a_1,...,$ $x<a_n$ (which we will write as $x<a_1,...,a_n$) then the denotation of x is given by

$$\phi(x) \subseteq U - \bigcup_i (\phi(a_i)), 1 \le i \le n.$$

That is, x denotes that set of individuals which is the whole universe without any individuals denoted by names that x is constrained from.

Further, we also need to ensure that any individual which is denoted by a term which contains a name which x is constrained from is not in $\phi(x)$. So, we can refine the expression given above to

$$\phi(x) = U - \{ \phi(t) \mid (\exists a)(a : t) \wedge (x < a)\}$$

where a : t is true iff the name a appears anywhere in the term t. We say that ϕ is closed with respect to constraints. Notice that we still have that if x<a then $\phi(x) \subseteq U - \phi(a)$.

The need for this final form of ϕ can be seen by considering, for example, the sentence S

$$(\exists x)(\forall y)(P(x,f(y)) \to P(x,x))$$

which is not valid. (Try the interpretation where P is <, f is successor and x is zero and y any natural number). However, without the closure of ϕ we can show that it is valid since $\neg S$ is unsatisfiable:

$$\neg(\exists x)(\forall y)(P(x,f(y)) \to P(x,x))$$
$$|$$
$$(\forall x)(\exists y)\neg(P(x,f(y)) \to P(x,x))$$
$$|$$
$$(\forall x)(\exists y)(P(x,f(y)) \wedge \neg P(x,x))$$
$$|$$
$$(\exists y)(P(x_1,f(y)) \wedge \neg P(x_1,x_1))$$
$$|$$
$$(P(x_1,f(a)) \wedge \neg P(x_1,x_1))$$
$$|$$
$$P(x_1,f(a))$$
$$|$$
$$\neg P(x_1,x_1)$$

which closes with x_1 replaced by f(a).

Definition 6.3. A *substitution* is a function
$$\sigma : S \to S$$
where S is the set of all sentences of our language.

If P is a sentence then Pσ is another possibly identical sentence. A substitution can be represented explicitly as a set of pairs
$$\{<d_1, t_1>,...,<d_n, t_n>\}$$
with d_i a dummy and t_i a term, $1 \le i \le n$, and we have

Definitions 6.4. each $<d_i, t_i>$ is called a *reduction* and if Σ is the set of all substitutions on S, i.e.
$$\Sigma \subseteq S \to S,$$
then if a given substitution σ is represented by a set of reductions such that each t_i is a ground term then we say that σ is a *ground substitution* and we write $\sigma \in \Sigma_0$.

For our purposes we need to realise that when a dummy has some other term substituted for it then the dummy and the term clearly have, henceforth, to have the same meaning. Thus, we see that substitutions modify the meaning of dummies. We need to ensure that a substitution is only allowed, we shall say 'proper', under certain a consistency condition on meaning.

Definition 6.5 A *proper substitution* is a substitution
$$\sigma = \{<x_1,t_1>,...,<x_n,t_n>\}$$
where
$$\phi(x_i) \cap \phi(t_i) \ne \{\}, 1 \le i \le n.$$

The condition expresses the fact that we only allow substitutions where a dummy's new meaning is consistent with its old one, where the consistency is decided by its constraints. It would not be sensible if after a substitution the new meaning of a dummy broke any constraints. Since the intended meaning of a dummy is the 'set of its possible meanings' we also clearly want the refinement of the denotation of a dummy after successive substitutions to be continuous in the sense that it does not 'jump' from one particular value to a totally unconnected one. It also ensures that we do not prejudge the new denotation of a dummy since this may lead to incompleteness. So, we have chosen the largest set which is consistent with the requirements as the new denotation.

For our model of dummies to accord with our intuitions we need to check that several properties hold in the model. Clearly, if a dummy x is such that x<a then we cannot allow the substitution $\{<x,a>\}$, i.e. it must not be proper. But if $\phi(x) \subseteq U - \phi(a)$ then the substitution is not proper according to the model since $\phi(x) \cap \phi(a) = \{\}$.

Further, if we have dummies x and y then we require that x and y inherit each others constraints and that they have the same denotation. As an example, we might have dummies x and y with x<a in
$$\neg A(x)$$
$$A(y)$$

and we clearly wish to do $\{<x,y>\}$. However, if later we also do $\{<y,a>\}$ then a constraint, though indirectly, has clearly been broken. So, y should inherit x's constraints. We can see that this should be so from realising that by symmetry we could also have had $\{<y,x>\}$. Then, x and y should now denote the same set and the place that was held by y in A(y) is now held by x as A(x), and so this place, and all others that y was in, are constrained from a.

Therefore, ϕ is modified by the substitution to ϕ' where $\phi'(x) = \phi'(y) = \phi(x) \cap \phi(y)$. Now we see that in the model the constraints on dummies are inherited in a uniform way. If x<a then $\phi(x) \subseteq U$ - $\phi(a)$ and now if we do $\{<x,y>\}$ then the new meaning of x is a subset of $(U - \phi(a)) \cap U$ from which it follows that $\phi'(x) = \phi'(y) \subseteq U$ - $\phi(a)$. That is, y<a and so y has inherited x's constraints as required. This clearly works in general.

If we have $\{<x,t>\}$ where t is a general term, i.e. not a name or a dummy, then it is clear that t should in some way inherit any constraints on x. For example,

$$(\exists x)(\forall y)(\exists z)(B(x,f(y)) \to (B(f(z),f(y)) \wedge B(x,f(z)))$$

is not valid. (Try f as successor, B as less than). However, we can form the following tableau

$$\neg(\exists x)(\forall y)(\exists z)(B(x,f(y)) \to (B(f(z),f(y)) \wedge B(x,f(z))))$$
$$|$$
$$(\forall x)(\exists y)(\forall z)\neg(B(x,f(y)) \to (B(f(z),f(y)) \wedge B(x,f(z))))$$
$$|$$
$$(\exists y)(\forall z)\neg(B(x_1,f(y)) \to (B(f(z),f(y)) \wedge B(x_1,f(z))))$$
$$|$$
$$(\forall z)\neg(B(x_1,f(a)) \to (B(f(z),f(a)) \wedge B(x_1,f(z)))) \qquad x_1 < a$$
$$|$$
$$\neg(B(x_1,f(a)) \to (B(f(x_2),f(a)) \wedge B(x_1,f(x_2))))$$
$$|$$
$$(B(x_1,f(a)) \wedge (\neg B(f(x_2),f(a)) \vee \neg B(x_1,f(x_2))))$$
$$|$$
$$B(x_1,f(a))$$
$$|$$
$$\neg B(f(x_2,f(a)) \vee \neg B(x_1,f(x_2))$$
$$\neg B(f(x_2,f(a))) \qquad \neg B(x_1,f(x_2))$$

and if we chose to do the right hand closure first by doing $\{<x_1,f(x_2)>\}$ and then the left hand by doing $\{<x_2,f(a)>\}$ we clearly have an unsound system. However, we need to ensure that after the first substitution that $f(x_2)<a$ which is inherited from $x_1<a$. Then, since constraints propagate to all terms which include the name a, we have $f(x_2)<f(a)$, i.e. $x_2<a$. Thus, the dummies in a term inherit the constraints of the dummy that they were substituted for. This is the case in general since if $f(t_1,...,t_n)<a$ then $f(t_1,...,t_n) < f(a,...,a)$ so if any t_i is, or contains, a dummy then it inherits the constraint. So, in the example above, after $\{<x_1,f(x_2)>\}$ we would have $x_2<a$ and so the second substitution would not be proper since

$$\phi(f(a)) \in \{\phi(s) \mid (\exists a)(a : s)(x_2<a)\}$$
so
$$\phi(x_2) \cap \phi(f(a)) = \{\}.$$
To sum up; we have that a proper substitution $\{<x,t>\}$ changes ϕ to ϕ' so that
$$\phi'(t) = \phi(x) \cap \phi(t)$$
and any dummies in t inherit all the constraints of x.

We can extend this to any proper substitutions:

Lemma 6.6 if σ is some proper substitution and t is some term then $\phi'(t\sigma) \subseteq \phi(t)$, where ϕ' is ϕ updated by any changes due to σ.

Proof by induction on the number of reductions in σ.

If $\sigma = \varepsilon$ then clearly $\phi' = \phi$ so $\phi'(t\sigma) \subseteq \phi(t)$. As an inductive hypothesis assume that if σ has k≥0 reductions then for any term t, $\phi'(t\sigma) \subseteq \phi(t)$.

Then, if σ has k+1 reductions then it can be factorized so that $\sigma = \sigma'\rho$ where ρ is a singleton. Since for any term u and substitutions θ, τ we have $u(\theta\tau) = (u \theta) \tau$ we can write $t\sigma = t(\sigma'\rho) = (t\sigma')\rho$. By hypothesis $t\sigma'$ is such that $\phi'(t\sigma') \subseteq \phi(t)$.

Now, ρ is a singleton (proper) substitution. We consider its effect on some term u. If ρ is of the form $\{<x,s>\}$ then if u does not contain x then $u\rho = u$. If u contains s then since $\phi'(s) \subseteq \phi(s)$ we have $\phi'(u) \subseteq \phi(u)$ and if not then $\phi'(u) = \phi(u)$. So, $\phi'(u\rho) \subseteq \phi(u)$.

If u does contain x then firstly, if u = x then $u\rho = x\rho = s$ and $\phi(u) = \phi(x)$ and $\phi'(s) \subseteq \phi(x)$ so $\phi'(u\rho) = \phi'(x\rho) = \phi'(s) \subseteq \phi(x) = \phi(u)$.

If u is of the form f(...,x,...), i.e. it contains x, then $u\rho$ is f(...,s,...) and since $\phi'(s) \subseteq \phi(x)$ then $\phi'(u\rho) \subseteq \phi(u)$. Then , by induction on the depths of the nesting of occurrences of x, we have that for any term u $\phi'(u\rho) \subseteq \phi(u)$.

Then, by induction on k, we have the required result.

Now, if we extend the usual negotiability condition on reductions to

Definition 6.7 A reduction $<x,t>$ is *negotiable* iff it satisfies

a) t does not contain the dummy x

b) t does not contain any name that x is restricted from

then we clearly have the following

Lemma (Negotiablity) 6.8. if all reductions in a substitution are negotiable then it is a proper substitution.

Definition 6.9. A tableau T_2 is a *direct extension* of a tableau T_1 iff it was derived from T_1 in any of the following ways:

D^*_1) as D_1 in 4.5

D^*_2) as D_2 in 4.5

D^*_3) if some γ occurs on a path p_l with leaf l then join $\gamma(x_i)$ to l, where x_i is some new dummy, as long as $\gamma(x_i)$ is not subsumed by some sentence already on p_l.

D^*_4) if some δ appears on p_l then join $\delta(a)$ to l, where a is some unused name. If any dummies appear in δ then record the constraint x<a for each dummy x in δ.

D^*_5) if two sentences of the form $P(s_1,...,s_n)$ and $\neg P(t_1,...,t_n)$ appear on some path and they have as mgu a proper substitution σ then perform σ on every sentence in T_1 to get T_2.

Lemma (direct extension) 6.10.

i) for any satisfiable set of sentences S the following hold:

S_1) as S_1 in 4.6

S_2) as S_2 in 4.6

S_3) if $\gamma \in$ S then S $\bigcup \{\gamma(t)\}$ is satisfiable for every term t appearing in S

S_4) as S_4 in 4.6

ii) if a tableau T_2 is a direct extension of tableau T_1 which is true under $<\phi,\psi>$ then T_2 is true under $<\phi',\psi>$, where ϕ' conservatively extends ϕ.

Proof the proof for (i) is a standard result as in [19] or [22].

ii) the cases for D_1^* to D_4^* are again standard. If T_2 was obtained from T_1 by the use of D_5^* then T_1 has had some (proper) substitution σ performed on each of its sentences. If S was any sentence of T_1 we consider its status in T_2, i.e. after σ was performed on it.

If Sσ=S then clearly Sσ and S have the same truth value. If S$\sigma \neq$ S then S must contain some dummy x such that $<x,t> \in \sigma$ for some term t.

If S = A($a_1,...,a_n$) is some sentence in T_1 then we have two cases to consider. If S was true in T_1 then $<\phi(a_1),...,\phi(a_n)> \in \psi(A)$ and since $\phi'(a_i\sigma) \subseteq \phi(a_i)$, $1 \leq i \leq n$, by lemma 6.6, we have
$$<\phi'(a_1s),...,\phi'(a_ns)> \in \psi(A)$$
so Sσ is true.

If $<\phi(a_1),...,\phi(a_n)> \notin \psi(A)$ then $<\phi'(a_1),...,\phi'(a_n)>\sigma \notin \psi(A)$ also, so Sσ is not true.

Thus, the truth value of S is maintained. By induction on the structure of more complicated sentences we have the required result.

Corollary 6.11. If the initial sentences of a tableau T are satisfiable then T is true.

Proof By repeated use of the direct extension lemma.

Soundness Theorem 6.12. Any provable entailment is valid

Proof Follows from the above.

Definition 6.13. The *constraint set*, $\kappa(x)$ of a dummy x is the set of all the terms that x is constrained from. It is calculated recursively on the structure of terms:

$$\kappa_0(x) = \{t \in T \mid x < t\}$$
$$\kappa_{i+1}(x) = \{t \in T \mid t = f(t_1,...,t_n) \text{ and } t_j \in \kappa_i(x), 1 \leq j \leq n\}$$
$$\kappa(x) = \bigcup_i \kappa_i(x)$$

Definition 6.14. A set of terms T *covers* a set of names N iff if $n \in$ N then either

1) $n \in$ T

2) $(\exists x \in$ T$)(x$ is a dummy and $n \notin \kappa(x))$

Definition 6.15. A set of sentences S is a *parameterized model set (pms)* iff it satisfies:

$P_0)$ if $A(s_1,...,s_n) \in$ S then there is no $\neg A(t_1,...,t_n)$ in S such that $A(s_1,...,s_n)$ and $A(t_1,...,t_n)$ have a proper mgu σ, and vice versa.

$P_1)$ as H_1 in 4.9

$P_2)$ as H_2 in 4.9

$P_3)$ if $\gamma \in$ S then there are sentences $\gamma(t_i) \in$ S such that $\bigcup_i \{t_i\}$ covers all the names that appear anywhere in S

$P_4)$ as H_4 in 4.9

Definition 6.16. A path p in a tableau T is *complete* iff

$CC_1)$ as (i) in 4.11

$CC_2)$ as (ii) in 4.11

$CC_3)$ for every γ on p, then there appear sentences $\gamma(t_i)$ on p such that $\bigcup_i \{t_i\}$ covers the set of all names that appear on p.

$CC_4)$ as (iv) 4.11

Definition 6.16. A tableau is *completed* iff all its paths are either closed or complete.

We show completeness by proving that if S is a pms then it is satisfiable and that we can complete a path by repeatedly directly extending it.

Lemma (Generalized Hintikka Lemma) 6.18. If S is a pms then S is satisfiable.

Proof Let I_U be an interpretation of the sentences in S. The universe U is the set of all the ground terms appearing in the sentences of S. We define ϕ as follows:

1) $\phi(a) = \{a\}$ for all names a

2) $\phi(x) = \{$all ground terms that can be substituted for x$\} = U - \kappa(x)$

3) $\phi(f) = \lambda s_1,...,s_n.f(s_1,...,s_n)$ for each n-ary function symbol f

Note. For instance, if $\phi(x_1) = \{a,b,c\}$ and $\phi(a) = \{a\}, \phi(b) = \{b\}, \phi(c) = \{c\}$ then

$\phi(g(a,h(x_1),b))$
$= \phi(g)(\phi(a),\phi(h(x_1)),\phi(b))$
$= \phi(g)(\{a\},\phi(h)(\phi(x_1)),\{b\})$
$= \phi(g)(\{a\},\phi(h)(\{a,b,c\}),\{b\})$

Now, $\phi(g) = \lambda s_1 s_2 s_3.g(s_1,s_2,s_3)$ and $\phi(h) = \lambda s.h(s)$ so
$$\phi(h)(\{a,b,c\}) = h(\{a,b,c\}) = \{h(e) \mid e \in \{a,b,c\}\} = \{h(a),h(b),h(c)\}$$
and

$\phi(g)(\{a\},\{h(a),h(b),h(c)\},\{b\})$
$= g(\{a\},\{h(a),h(b),h(c)\},\{b\})$
$= \{g(e_1,e_2,e_3) \mid e_1 \in \{a\} \wedge e_2 \in \{h(a),h(b),h(c)\} \wedge e_3 \in \{b\}\}$
$= \{g(a,h(a),b), g(a,h(b),b), g(a,h(c),b)\}$

End of note.

Now we need to define ψ. We use induction on the structure of sentences in S.

1) if $P(t_1,...,t_n) \in$ S then $P(t_1,...,t_n)$ is true by definition

2) if $\neg P(t_1,...,t_n) \in$ S then there are no $P(s_{1j},...,s_{nj})$ in S such that $P(t_1,...,t_n)$ and $P(s_{1j},...,s_{nj})$ have a proper mgu, by P_0. Now, assume that $P(t_1,...,t_n)$ is true, so that $<\phi(t_1),...,\phi(t_n)> \in \psi(P)$, i.e. $\{(e_1,...,e_n) \mid e_i \in \phi(t_i)\} \subseteq \psi(P)$. Since $\psi(P)$ is determined by all the sentences in S of the form $P(u_1,...,u_n)$, $\psi(P) = \bigcup_j \{(e_1,...,e_n) \mid e_i \in \phi(u_{ij})\}$ for each $P(u_{1j},...,u_{nj}) \in$ S so we have that

$$\{(e_1,...,e_n) \mid e_i \in \phi(t_i)\} \subseteq \bigcup_j \{(e_1,...,e_n) \mid e_i \in \phi(u_{ij})\}.$$

Now, let $\tau_1 \in \Sigma_0$. Then, since $\phi'(t_i\tau_1) \subseteq \phi(t_i)$, $P(t_1,...,t_n)\tau_1$ is still true. Further, $\psi(P)$ is now determined by all the $P(u_{1j},...,u_{nj})\tau_1$ so we have

$$\{(e_1,...,e_n) \mid e_i \in \phi'(\tau_1)\} \subseteq \bigcup_j \{(e_1,...,e_n) \mid e_i \in \phi'(u_{ij}\tau_1)\}.$$

We will write this as

$$<\phi'(t_i\tau_1)> \subseteq \bigcup_j <\phi'(u_{ij}\tau_1)>$$

Clearly, we can repeat this process with τ_2, τ_3, with the relevant version of the relation above holding in each case, until $<\phi^{(m)}(t_i\tau_1...\tau_m)>$ is a singleton , i.e. we have circumscribed every dummy in $P(t_1,...,t_n)$ to stand for a definite (singleton) individual. At this stage at least one of the $<\phi^{(m)}(s_{ij}\tau_1...\tau_m)>$ must contain this singleton since

$$<\phi^{(m)}(t_i\tau_1...\tau_m)> \subseteq \bigcup_j <\phi^{(m)}(u_{ij}\tau_1...\tau_m)>$$

Now, we can choose $\rho \in \Sigma_0$ to be a proper substitution such that $<\phi^{(m)}(s_{ij}\tau_1...\tau_m\rho)>$ is the same singleton. This means that there is a k such that for each i

$$s_{kj}\tau_1...\tau_m\rho = t_i\tau_1...\tau_m$$

but since the right hand side of this equality is ground we also have

$$s_{ij}\tau_1...\tau_m\rho = t_i\tau_1...\tau_m\rho = \theta$$

so for at least one j, $P(s_{1j},...,s_{nj})$ and $P(t_1,...,t_n)$ have a proper unifier θ, so they also have a proper mgu. However, this is a contradiction, so $<\phi(t_1),...,\phi(t_n)> \notin \psi(P)$, so $P(t_1,...,t_n)$ is not true, so $\neg P(t_1,...,t_n)$ is true.

3) for a sentence of type α, by the standard method

4) for a sentence of type β, by the standard method

5) if $\gamma \in$ S then there are some dummies x_i such that $\gamma(x_i) \in$ S and the set of all the x_i covers the ground terms in S by P_3. By hypothesis the $\gamma(x_i)$ are all true. Since the x_i cover all the ground terms in S it follows that every individual in U is in the denotation set of at least one dummy. So, $\gamma(a)$ is true for every name in S so γ is true by T_3.

6) for a sentence of type δ, by the standard method

Thus S has a model and so is satisfiable.

Clearly, any completed path in a tableau is such that its sentences form a pms, so we have, assuming we have an algorithm that completes a path if at all possible

Completeness Theorem 6.19. If T $\models$ S is valid then it is provable.

Corollary 6.20. If T $\models$ S is valid then it is provable in a finite number of steps.

We now have the same problem as before of finding some way of repeatedly directly extending the tableau formed from the initial sentences in a way that is sure to lead to a completed tableau if one exists. In fact, we can, largely, follow our previous algorithm since, for the same reasons as given before, it is adequate. However, we do need to generalise the closing part to include a search for sentences which are contradictory due to a unifying substitution.

This may cause a problem, in general, since two different pairs of sentences are likely to require inconsistent substitutions, e.g. we may require both $\{<x_1,a>\}$ and $\{<x_1,b>\}$. Clearly, we cannot carry out both of them but if something is not done about this then the algorithm will not be complete. The solution is to assume that we have available enough 'copies' of the sentences in which the dummies concerned first appeared to allow, in some way, the results of both substitutions to be expressed.

An implementation of such a correct algorithm is given in [19]. However, we can circumvent the problem of keepimg track of and resolving clashes if we skolemize our sentences so that no existential quantifiers appear. This is clearly adequate for a theorem-proving implementation, but does somewhat diminish the properties of readability, comprehensibility etc. that we were originally motivated by. However, on the positive side, it can lead to somewhat shorter proofs. The table below shows some results comparing the two alternatives, i.e. using the implementation of the 'full' system where possibly inconsistent substitutions are handled and by skolemizing, by listing the number of nodes required to generate a proof. The sentences used in this comparison are mainly drawn from [10]. They are as follows:

Ex1	$(\exists x)(\forall y)\{[((A(y)\to B(y))\leftrightarrow A(x)) \wedge ((A(y)\to D(y))\leftrightarrow B(x)) \wedge$ $(((A(y)\to B(y))\to D(y))\leftrightarrow D(x))] \to (\forall z)(A(z)\wedge B(z)\wedge D(z))\}$
Ex2	$\{[(\forall x)(\exists y)(A(x,y) \vee A(y,x)) \wedge (\forall x)(\forall y)(A(x,y)\to A(y,y))] \to (\exists z)A(z,z)\}$
Ex3	$(\exists x)(\forall y)(\forall z)\{[((A(y,z)\to(B(y)\to D(x)))\to A(x,x)) \wedge$ $((A(z,x)\to B(x))\to D(z)) \wedge A(x,y)] \to A(z,z)\}$
Ex4	$(\forall x)(\exists y)\{[(\exists u)(\forall v)(A(u,x)\to (B(v,u) \wedge B(u,x)))] \to$ $([(\exists u)(\forall v)(A(u,y)\to(B(v,u) \wedge B(u,y)))] \vee$ $[(\forall u)(\forall v)(\exists w)((B(v,u) \vee D(w,y,v))\to B(u,w))])\}$
Ex5	$\{[(\forall x)(K(x)\to(\exists y)(L(y) \wedge (A(x,y)\to B(x,y)))) \wedge (\exists z)(K(z) \wedge (\forall u)(L(u)\to A(z,u)))]$ $\to (\exists v)(\exists w)(K(v) \wedge L(w) \wedge B(v,w))\}$
Ex6	$(\exists x)(\exists y)(\forall z)\{ (A(x,y)\to(A(y,z) \wedge A(z,z))) \wedge ((A(x,y) \wedge B(x,y))\to$ $(B(x,z) \wedge B(z,z))) \}$

and the results were

	original	skolemized
Ex1	815	783
Ex2	54	21
Ex3	370	178
Ex4	96	24
Ex5	71	50
Ex6	no proof	105

7. A bibliography: semantic tableaux in theorem-proving

In this section we present references to as much on the work done in theorem-proving using semantic tableaux as space and our current awareness allows. As can be seen, since workers in the field have tended to be rather isolated, many ideas recur and many techniques are re-invented.

7.1. Beth 1955 [2] and 1958 [3].

In each of these papers, the basic method of semantic tableaux is introduced. We also see the appearance of the idea of constructing proofs from refutations, and explicit mention of a *machine* to do the theorem-proving.

7.2. Jeffrey 1967 [12].

This is an excellent introductory logic textbook which presents the task of proving the validity of arguments in an algorithmic way. It also extends the method to first-order theories with functions and equality, though in an unsophisticated way from the point of view of feasible computations.

7.3. Popplestone 1967 [16].

This is the first example of using more sophisticated, 'tailor-made' techniques for dealing with equality in semantic tableaux.

7.4. Cohen et al. 1974 [8].

This paper is partly an exercise in using ALGOL-68, but is also the first to explicitly mention using unification to cut down the size of the search-space.

7.5. Broda 1980 [7].

This paper is the first presentation of extended examples of incorporating unification into semantic tableaux, mainly in the context of logic programming. It shows how many of the different forms of resolution are expressible as heuristics that control the order of applying the different rules.

It also discusses at length some resolution based techniques in a unified semantic tableau setting, and therefore allows some comparison of various techniques.

7.6. Reeves 1983 [18].

This is an introductory treatment of semantic tableaux using unification in a notation introduced by Smullyan [22] and much used since. It presents some (rather abstract) algorithms too.

7.7. Wrightson 1984 [24].

This paper also presents the use of unification in a semantic tableau setting, but also adapts the connection ideas of Andrews [1], Bibel [4] and Kowalski [13].

7.8. Schönfeld 1985 [21].

Here the author introduces a tableau-based system with the intention of viewing it as an interpreter for full first-order logic programs, but using the PROLOG search strategy and extending it to the full language case.

7.9. Wallen 1985 [23].

This presents a way of transforming a tableau proof system into a connection based system with the aim of enhancing efficiency and also extends the method to deal with other than classical logics.

7.10. Oppacher and Suen 1986 [15].

Discussion of many heuristics for enhancing the efficiency of semantic tableau theorem-provers. It also contains a very elegant and persuasive piece on the use of semantic tableaux as a basis for automatic theorem-proving.

8. Non-standard logics in a semantic tableau framework.

The text by Fitting [9] gives a full account of work using semantic tableaux to present proof systems for intuitionistic and modal logics, based on the work of Kripke. Fitting also presents algorithms for the proof systems that he treats.

Rescher and Urquhart [20], among other things, do the same for temporal logics.

In all these cases what we might term 'abstract' algorithms are presented (rather as in the first part of this paper) in that they do not address the problems of actually implementing efficient algorithms which would lead to proofs by feasible computations (and of course that was not their aim). However, all the background theory is there, and so, when combined with the efficiency enhancing work referred to in section 7, they provide the basis of much future activity in automatic theorem-proving in non-standard logics.

References

1. Andrews, P., *Theorem Proving via General Matings*, JACM, vol. 28, no. 2, 1981.
2. Beth, E. W., *Semantic Entailment and Formal Derivability*, 1955 in Hintikka, Philosophy of Mathematics, Oxford, 1969.
3. Beth, E.W., *On Machines which Prove Theorems*, 1958, collected in *Automation of Reasoning 1*, ed. J. Siekmann and G. Wrightson, Springer-Verlag, 1983.
4. Bibel, W., *Matings in Matrices*, CACM, vol. 26, no. 11, 1983, pp. 844-852.
5. Bibel, W., Schreiber J., *Proof search in a Gentzen-like system of First-Order Logic*, International Computing Symposium, 1975.
6. Bishop,E., *Foundations of Constructive Mathematics*, McGraw-Hill,1966.
7. Broda K., *The relation between semantic tableaux and resolution theorem-provers*, Internal report, Department of Computing and Control, Imperial College, University of London, 1980.

8. Cohen J., Trilling L., Wegner P., *A Nucleus of a Theorem-Prover Described in ALGOL-68*, International Journal of Computer and Information Sciences, vol.3 no.1, 1974.

9. Fitting, M.C., *Proof Methods for Modal and Intuitionistic Logics*, Reidel, Dordrecht, The Netherlands, 1983.

10. Gilmore, P.C., *A Proof Method for Quantification Theory: Its Justification and Realization*, I.B.M. Journal of Research and Development 4, 1960.

11. Hintikka, J., *Form and Content in Quantification Theory*, Acta Philosophica Fennica, VIII, 1955.

12. Jeffrey R. C., *Formal Logic: Its Scope and Limits*, McGraw-Hill, 1966.

13. Kowalski, R., *A Proof Procedure using Connection Graphs*, JACM, vol. 22, no. 4, 1975.

14. Martin-Löf, P., *Constructive Mathematics and Computer Science*, Department of Mathematics, University of Stockholm and 6th. International Congress of Logic and Methodology of Science, Hannover 1979.

15. Oppacher F., Suen E., *Controlling deduction with proof condensation and heuristics*, Proc. CADE-8, LNCS 230, pp.384-393, 1986.

16. Popplestone, R.J., *Beth Tree Methods in Automatic Theorem-Proving*, Machine Intelligence 1, Oliver and Boyd, Edinburgh,1966.

17. Prawitz, D., *An Improved Proof Procedure, Theoria, 26, 1960*.

18. Reeves, S.V., *An Introduction to Semantic Tableaux*, Department of Computer Science, CSM-55, University of Essex, 1983.

19. Reeves, S.V., *Theorem-proving by Semantic Tableaux*, Ph.D. thesis, University of Birmingham, 1985.

20. Rescher, N. and Urquhart, A., *Temporal Logic*, Springer-Verlag, 1971.

21. Schönfeld W., *Prolog extensions based on tableau calculus*, Proc. IJCAI-85, pp.730-732, 1985.

22. Smullyan, R.M., *First-Order Logic*, Ergebnisse der Mathematik und ihrer Grenzgebiete, 43, Springer-Verlag, Berlin, 1968.

23. Wallen, L., *Generating Connection Calculi form Tableaux and Sequent Based Systems*, Proc. AISB 1985.

24. Wrightson G., *Semantic Tableaux, Unification and Links*, Technical Report CSD-ANZARP-84-001, Department of Computer Science, University of Wellington, Victoria, New Zealand, 1984.

A Computationally Efficient Proof System for S5 Modal Logic

Lincoln A. Wallen   Gregory V. Wilson
Department of Artificial Intelligence
Edinburgh University
Scotland

Abstract

We present a computationally efficient matrix proof system for S5 modal logic. The system requires no normal-form and admits a natural implementation using structure-sharing techniques. In addition, proof search may be interpreted as constructing generalised proofs in an appropriate sequent calculus, thus facilitating its use within interactive environments. We describe features of an implementation developed from an existing implementation of a matrix proof system for first-order logic.

1 Introduction.

Modal logics are widely used in various branches of artificial intelligence and computer science as logics of knowledge and belief (*eg.*, [Moo80,HM85,Kon84]), logics of programs (*eg.*, [Pne77]), and for specifying distributed and concurrent systems (*eg.*, [HM84,Sti85]). As a consequence, the need arises for proof systems for these logics which facilitate efficient automated proof search.

The main hurdle to the application of resolution based techniques to non-standard logics is that the techniques are formulated under the assumption that the input formulae are in clausal form [CL73]. Most non-standard logics of interest fail to admit such a normal-form.

Bibel's connection calculus [Bib81,Bib82a] is a non-clausal proof system for first-order logic comparable in computational efficiency to the most efficient of the clausal techniques for that logic [Bib82b]. In [Wal86] it was shown that far from being an ad-hoc proof system for one particular logic, the connection calculus could be seen as a framework for implementing sequent- and tableau-based proof systems in a computationally efficient manner. Since most of the non-standard logics of interest admit such proof systems we can make use of this analysis to develop connection calculi for them. This paper fulfills an undertaking to present the details of such an application to S5 modal logic.

We begin by presenting Kanger's sequent calculus for S5 [Kan57] using a notation developed by Smullyan and Fitting [Smu68,Fit72]. Next, we develop a connection calculus from the sequent system using the techniques discussed in [Wal86]. In addition to the basic theory, we give details of an implementation developed directly from an existing implementation of the connection calculus for standard first-order logic. The adaptation from first-order classical logic to S5 proved quite straightforward. Further details of this implementation can be found in [Wil86].

Because we do not need to give up our sequent interpretation of the proof system, it is possible to view proof search within the modal connection calculus as a process which constructs a form of sequent proof tree. Brief details of this are given. Such a facility encourages the use of these techniques within interactive environments (*cf.* [BT75]).

141

2   Preliminaries.

2.1   Syntax, semantics and notation.

Modal *formulae* are defined as usual by adding the formation rule:

if A is a formula, then so are $\Box\, A$ and $\Diamond A$,

to the formation rules for propositional formulae. We let A, B, C range over modal formulae. A *signed* modal formula is a pair $\langle A, n \rangle$, where A is a formula and $n \in \{0, 1\}$. We let X, Y, Z range over signed modal formulae.

Following Smullyan [Smu68] and Fitting [Fit83] we classify signed modal formulae and their principal signed subformulae as follows:

α	α_1	α_2	β	β_1	β_2	ν	ν_0
$\langle A \wedge B, 1 \rangle$	$\langle A, 1 \rangle$	$\langle B, 1 \rangle$	$\langle A \wedge B, 0 \rangle$	$\langle A, 0 \rangle$	$\langle B, 0 \rangle$	$\langle \Box A, 1 \rangle$	$\langle A, 1 \rangle$
$\langle A \vee B, 0 \rangle$	$\langle A, 0 \rangle$	$\langle B, 0 \rangle$	$\langle A \vee B, 1 \rangle$	$\langle A, 1 \rangle$	$\langle B, 1 \rangle$	$\langle \Diamond A, 0 \rangle$	$\langle A, 0 \rangle$
$\langle A \Rightarrow B, 0 \rangle$	$\langle A, 1 \rangle$	$\langle B, 0 \rangle$	$\langle A \Rightarrow B, 1 \rangle$	$\langle A, 0 \rangle$	$\langle B, 1 \rangle$	π	π_0
$\langle \neg A, 1 \rangle$	$\langle A, 1 \rangle$	$\langle A, 1 \rangle$				$\langle \Box A, 0 \rangle$	$\langle A, 0 \rangle$
$\langle \neg A, 0 \rangle$	$\langle A, 0 \rangle$	$\langle A, 0 \rangle$				$\langle \Diamond A, 1 \rangle$	$\langle A, 1 \rangle$

We shall use $\alpha, \alpha_1, \alpha_2, \beta, \beta_1, \beta_2, \ldots$ to denote signed formulae and their components of the respective types.

For example, a signed formula of the form $\langle A \wedge B, 1 \rangle$ is an α according to the first table above. Its components $\langle A, 1 \rangle$ and $\langle B, 1 \rangle$ are denoted by α_1 and α_2 respectively.

An S5-*model structure* is a pair $\langle G, R \rangle$ consisting of a set G and an equivalence relation R on G. An S5-model is a triple $\langle G, R, \Vdash \rangle$ where $\langle G, R \rangle$ is an S5-model structure, and $\Vdash$ is a relation between elements of G and signed modal formulae which, for all $w \in G$ satisfies:

1. exactly one of $w \Vdash \langle A, 1 \rangle$ or $w \Vdash \langle A, 0 \rangle$;

2. $w \Vdash \alpha$ iff $w \Vdash \alpha_1$ and $w \Vdash \alpha_2$;

3. $w \Vdash \beta$ iff $w \Vdash \beta_1$ or $w \Vdash \beta_2$;

4. $w \Vdash \nu$ iff for all $v \in G$ such that wRv, $v \Vdash \nu_0$;

5. $w \Vdash \pi$ iff for some $v \in G$ such that wRv, $v \Vdash \pi_0$.

A formula A is S5-*valid* in the S5-model $\langle G, R, \Vdash \rangle$ iff for all $w \in G$, $w \Vdash \langle A, 1 \rangle$. A formula is S5-*valid* iff it is S5-valid in all S5-models.

2.2   A sequent calculus for S5.

Let $\langle G_0, R_0 \rangle$ be an S5-model structure, fixed for the rest of this section. We use p, q to denote elements of G_0. We refer to these elements as *prefixes*. For $p \in G_0$, pX is called a *prefixed signed formula*. Prefixes are used to name possible worlds in some arbitrary model. R_0 is used to represent the relation of accessibility between prefixes and hence between possible worlds in that model. pX is satisfied by a model just when the world denoted by p, say $\iota(p)$, satisfies X; *i.e.*, $\iota(p) \Vdash X$.

For our purposes *sequents* are merely sets of prefixed signed formulae. We use S, pX to denote $S \cup \{ pX \}$. Note that pX may occur in S.

We now present Kanger's system for S5 using this notation. The basic sequent is a sequent which contains both $p\langle A, 1 \rangle$ and $p\langle A, 0 \rangle$ for some atomic formula A, *i.e.*,

$$S, p\langle A, 1 \rangle, p\langle A, 0 \rangle.$$

The operational rules are stated concisely thus:

$$\frac{S,\, p\alpha_1, p\alpha_2}{S,\, p\alpha}\ \alpha \qquad\qquad \frac{S,\, p\beta_1 \quad S,\, p\beta_2}{S,\, p\beta}\ \beta$$

$$\frac{S,\, q\nu_0}{S,\, p\nu}\ \nu \qquad\qquad \frac{S,\, q\pi_0}{S,\, p\pi}\ \pi$$

The π-rule is subject to the following proviso:

q must not occur in S (*i.e.*, q must not prefix any other formula in S).

Derivations and *proofs* are defined as usual. A formula A is a *theorem* if there is a proof of the sequent $\{\, p\langle A, 0\rangle \,\}$ for some (arbitrary) prefix p. This calculus is both sound and complete for S5 [Kan57].

Remark. The usual representation of sequents can be recovered by introducing a *sequent arrow* $\longrightarrow$ and placing those prefixed formulae signed 1 on the left of the arrow and those signed 0 on the right.

In practice we use the rules from conclusion to premises; *i.e.*, we start with the sequent $\{\, p\langle A, 0\rangle \,\}$ and build a proof tree backwards from this root to the leaves. Used in this way, the sequent system is equivalent to Fitting's prefixed tableau system for S5 [Fit72]. The force of the proviso on the π-rule is that the prefix q must be completely new to the upper sequent. ∎

3   A connection calculus for S5.

As discussed in [Wal86] the proviso on the π rule introduces an order dependence in the search for a proof. If a prefix is introduced by the use of the ν rule it cannot subsequently be introduced by the use of the π rule. The path to an basic sequent may therefore be blocked by injudicious choices of the prefix q in both (inverted) modal rules. This fact, together with the blind *connective-driven* search for the appropriate pairs of atoms to form a basic sequent, makes direct implementation of the sequent system inefficient. In [Wal86] we show how inefficiencies such as these are tackled in the case of first-order logic by Bibel's connection calculus.

In the context of modal logic the key techniques are:

- *connection driven search:* putative instances of the basic sequent are identified using structural properties of the formula.

- *delayed choice of prefix:* the symbols introduced as prefixes during use of the ν rule are considered as "variables." The prefixes introduced by applications of the π rule are considered as "constants." The instantiation of the variable prefixes is driven by the choice of connection (basic sequent) which requires the prefixes of the distinguished atomic formulae to be identical.

- *reduction ordering:* the substitution mentioned above is considered admissible if a correct (partial) proof tree can always be constructed from the current set of connections. In particular this means respecting the proviso on the π-rule. This property of the substitution is checked directly rather than by actually constructing such a proof tree.

- *structure sharing:* the intermediate states of the proof are encoded using pointers into the original formula. No new formulae need be considered. In particular, multiple instances of subformulae (in this case instances of formulae dominated by modal operators) are obtained by an indexing method rather than explicit copying.

3.1 Formula trees.

A *formula tree* for a signed modal formula is a variant of its formation tree containing additional information as to the polarity of its subformulae. It is best explained by example. Figure 1 shows the formula tree for the formula $\langle \Box((A \wedge B) \wedge \Diamond C \Rightarrow \Diamond(A \wedge C)), 0 \rangle$. The piece of

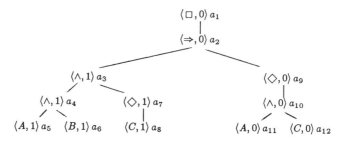

Figure 1: A formula tree.

concrete syntax associated with a node or *position* in the tree is called the *label* of that position.

We classify the positions of a formula tree according to the classification of the subformula rooted at that position. Thus if the subformula is an α, then the position is an α; if it is a β, the position is a β, etc. Additionally, we classify the root node of the formula tree as a π_0.

Notice that each position has two types: its principal type $(eg., \alpha, \beta, \nu, \ldots)$ is determined by its label and polarity, while its secondary type $(eg., \alpha_1, \alpha_2, \beta_1, \ldots)$ arises from the type of its parent.

For a given formula tree we use k, l, possibly subscripted, to denote positions and V_0 and Π_0 denote the sets of positions of type ν_0 and π_0 respectively.

3.2 Modal multiplicity.

The inverted rule enables any prefix q to be introduced into a sequent to prefix formulae rooted at ν_0-type positions. Within the context of a sequent proof this is the mechanism by which one works towards instances of the basic sequent to complete a branch.

The definitions of this section are introduced for a given formula tree for a given signed formula X. A function μ_M from V_0 to the positive integers is called a *modal multiplicity* for X. A modal multiplicity serves to encode the number of distinct instances of ν_0-type (sub)formulae used within a putative proof.

If μ_M is a modal multiplicity for X we define the (indexed) formula tree for the *indexed formula* X^{μ_M} as a tree of indexed positions of the form k^κ, where k is a position of the basic formula tree for X and κ is a sequence of positive integers defined as follows: if $k_1 < k_2 < \cdots < k_n \leq k$, $0 \leq n$, are those ν_0-type positions that dominate k in the basic formula tree for X, then

$$\kappa \in \{ (j_1 j_2 \cdots j_n) \mid 1 \leq j_i \leq \mu(k_i), \ 1 \leq i \leq n \}.$$

The ordering in the indexed tree $<^{\mu_M}$ is defined in terms of the ordering on the underlying tree. For indexed positions k^κ and l^τ

$$k^\kappa <^{\mu_M} l^\tau \quad \text{iff} \quad k < l \quad \text{and} \quad \tau = \kappa\theta,$$

where θ is some sequence of positive integers. The polarity and label of an indexed position k^κ is taken to be the same as the polarity and label of its underlying position. Consequently, indexed positions inherit the type of their underlying position.

We let u, v, possibly subscripted, range over indexed positions when we are not interested in the index, and omit the superscript on $<$. We abuse our notation and let ν_0, Π_0, etc., denote the sets of indexed positions of an indexed formula tree of the appropriate types. Henceforth we shall refer to indexed positions simply as positions.

Figure 2 shows the indexed formula tree for our example formula with a modal multiplicity of $\mu_M(a_{10}) = 2$ and 1 otherwise. As a convention we omit empty indices.

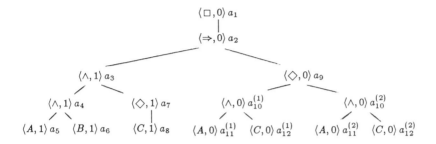

Figure 2: An indexed formula tree.

3.3 Paths and connections.

A *path through* X^{μ_M} is a subset of the positions of its formula tree. We shall use s, t, possibly subscripted, to denote these paths, and adopt the notation $s[u]$ for a path s with an occurrence of the position u. The set of paths through X^{μ_M}, is the smallest set such that:

1. $\left\{ k_0^{()} \right\}$ is a path, where $k_0^{()}$ is the root node of the formula tree for X^{μ_M};

2. if $s[\alpha^\kappa]$ is a path, so is $(s - \{\alpha^\kappa\}) \cup \{\alpha_1{}^\kappa, \alpha_2{}^\kappa\}$;

3. if $s[\beta^\kappa]$ is a path, so are $(s - \{\beta^\kappa\}) \cup \{\beta_1{}^\kappa\}$ and $(s - \{\beta^\kappa\}) \cup \{\beta_2{}^\kappa\}$;

4. if $s[\nu^\kappa]$ is a path, so is $s \cup \{\nu_0{}^{\kappa j}\}$, $1 \le j \le \mu_M(\nu_0)$;

5. if $s[\pi^\kappa]$ is a path, so is $s \cup \{\pi_0{}^\kappa\}$.

The path $(s - \{\alpha^\kappa\}) \cup \{\alpha_1{}^\kappa, \alpha_2{}^\kappa\}$ is said to have been *obtained by reduction on* α^κ from s; similar terminology is used in the other cases. Each path s through X^{μ_M} determines a set (or *sequent*) of positions as follows:

$$S(s) = \{ x \mid x \le y \quad \text{for some} \quad y \in s \}.$$

A path s through X^{μ_M} is *atomic* iff for every k^κ in s either

(a) k is labeled by an atomic formula, or

(b) k is a ν and for all j, $1 \le j \le \mu_M(\nu_0)$, $\nu_0{}^{\kappa j} \in S(s)$.

Remark. Our definition of path differs from Andrews' [And81] and Bibel's [Bib81] definition so as to demonstrate the relationship between the matrix methods and sequent/tableau methods. Their paths correspond roughly to our atomic paths. Each clause in our definition, when interpreted as operating on the sequent associated with the path, corresponds to an (inverted) rule of Kanger's system. Furthermore, for a given multiplicity, the sequent associated with an atomic path is *complete* in the sense that its membership cannot be increased by the application of further rules. These relationships are discussed in more detail in [Wal86]. ∎

We can see the atomic paths through an indexed formula by writing the components of an α-type subformula side by side and the components of a β-type subformula one above the other to form a nested *matrix*. Multiple instances of ν_0-subformulae are treated as multiple components of an α-type parent. The matrix representation of the indexed formula of Figure 2 is shown in Figure 3. The atomic paths through the formula are then (roughly) the horizontal

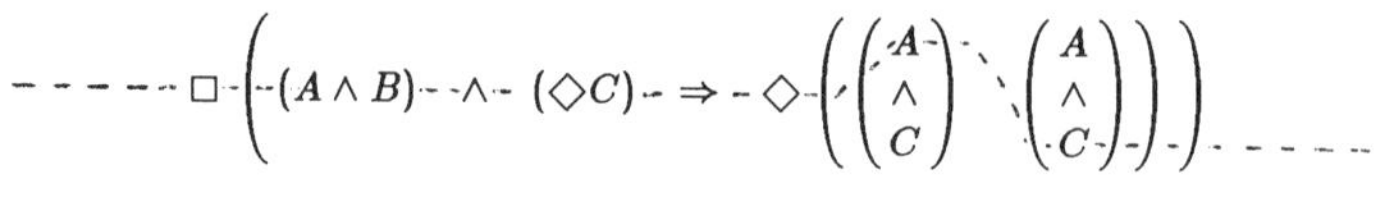

$$- - - - - \Box \cdot \left(\cdot (A \wedge B) \cdot \cdot \wedge \cdot (\Diamond C) \cdot \Rightarrow \cdot \Diamond \cdot \left(\cdot \left(\begin{matrix} A \\ \wedge \\ C \end{matrix} \right) \cdot \cdot \left(\begin{matrix} A \\ \wedge \\ C \end{matrix} \right) \right) \right) \cdot - - - - $$

Figure 3: Matrix representation of a formula.

matrix paths that consist of atomic formulae. There are four such paths in our example. One is shown as a dotted line in the figure.

A *connection* in X^{μ_M} is an unordered pair of positions of its formula tree labeled by the *same* atomic formula but of *different* polarities. In terms of the sequent system we wish to interpret a connection as an instance of a basic sequent. A set of connections is said to *span* X^{μ_M} just when every atomic path through X^{μ_M} contains a connection from the set. For example, the connections $\left\{ a_5, a_{11}^{(1)} \right\}$ and $\left\{ a_8, a_{12}^{(1)} \right\}$ span our example formula.

3.4 Complementarity.

In order to interpret a connection as an instance of a basic sequent we must ensure that the two atomic formulae represented by the positions of a connection have the same prefix. These prefixes are determined when the modal operator dominating the atomic formula is reduced using the inverted modal rules.

Let T denote the set $\nu_0 \cup \Pi_0$. We use pre(u) to denote the $<$-greatest T-element that dominates u in the formula ordering. pre(u) is called the *prefix* of u. Note that since the root node of a formula tree is a T-element this notion is well-defined. We shall use p, q, to denote positions when we are considering them as prefixes. In our ongoing example, the prefix of a_5 is a_2, whereas the prefix of $a_{11}^{(1)}$ is $a_{10}^{(1)}$.

The next step is to notice that we are free to choose the prefix introduced when we use the ν-rule, but must introduce arbitrary new prefixes when utilising the π-rule. We have indicated that the two positions that constitute a connection must have the same prefix. Consequently we treat ν_0-type prefixes as variables and π_0-type prefixes as constants and build a *modal substitution* $\sigma: \nu_0 \rightarrow T$ under which the required prefixes are identical.

For example, the modal substitution $\sigma = \left\{ a_{10}^{(1)} \leftarrow a_2 \right\}$ renders the connection $\left\{ a_5, a_{11}^{(1)} \right\}$ complementary. Recall that a_2 is a π_0-type positon and hence a constant whereas $a_{10}^{(1)}$ is a ν_0-type position and therefore a variable. We cannot build a consistent substitution that also makes the connection $\left\{ a_8, a_{12}^{(1)} \right\}$ complementary since this would have to involve the component $a_{10}^{(1)} \leftarrow a_8$ and a_2 and a_8 are distinct constants.

The substitution of a π_0 prefix for a ν_0 prefix entails that the former is introduced in place off the latter with the ν-rule. But this means that π_0 prefixes may be introduced into a sequent before the use of the π rule that introduced them from their parent and hence the proviso on the rule would not be met. We must ensure that this never happens.

A modal substitution $\sigma: \nu_0 \rightarrow T$ induces an equivalence relation $\sim_M$ and a relation $\sqsubseteq_M$ on $T \times T$ as follows:

1. If $\sigma(u) = v$ for some v of ν_0-type, then $u \sim_M v$.

2. If $\sigma(u) = v$ for some v of π_0-type, then $v \sqsubseteq_M u$.

3. If $v \sqsubset_M u$ and $u \sim_M u'$, then $v \sqsubset_M u'$.

The substitution σ is S5-*admissible* provided the reduction ordering $\lhd \overset{\text{def}}{=} (< \cup \sqsubset_M)^+$ is irreflexive.

The relation $v \sqsubset_M u$ between a π_0 and a ν_0 position indicates that the formula rooted at the parent of v (a π-type formula) should be reduced using the π rule to introduce the prefix v before the parent of u (a ν-type formula) is reduced using the ν rule to introduce the prefix u, the value of which is also v under the modal substitution. The equivalence relation $\sim_M$ indicates that the two ν_0-type prefixes must take the same value under the substitution.

Let σ be an S5-admissible mapping for X^{μ_M}. A connection $\{x, y\}$ in X^{μ_M} is said to be σ-*complementary* iff $\sigma(\text{pre}(x)) = \sigma(\text{pre}(y))$. A set of connections is said to be σ-complementary iff all its elements are σ-complementary.

We are now in a position to state our extension of Bibel's connection theorem to S5 modal logic.

Theorem 3.4.1 *A formula A is* S5-*valid iff there is a modal multiplicity* μ_M, *an* S5-*admissible mapping* σ *and a set of* σ-*complementary connections that spans* $\langle A, 0 \rangle^{\mu_M}$.

We omit the proof of this theorem due to lack of space. Correctness follows directly from the correctness of Kanger's sequent system together with the argument that the irreflexivity of $\lhd$ ensures the existence of a correct sequent proof respecting the proviso on the π rule. Completeness is obtained by showing that a modal multiplicity can be constructed such that the sequent associated with an atomic path containing no complementary connection forms an S5-Hintikka set which is realizable (see [Fit83]).

4 Implementation.

An implementation of this connection calculus for S5 has been developed from an existing implementation of a connection calculus for first-order logic [Wal83]. The implementation language chosen was Quintus PROLOG since we were more concerned with the techniques used to implement the proof system than in absolute efficiency. This section describes some of the interesting features of the implementation, a more comprehensive description can be found in [Wil86].

4.1 Static data structures.

4.1.1 Formula tree representation.

Each position of the formula tree must be stored in some space-efficient way which permits a time-efficient lookup mechanism. It is important that this storage method also allow the program to manipulate portions of the formula tree without actually making copies of those portions, so that demands on storage space do not become excessive. The formula tree forms the basis for an implementation utilising structure sharing.

For both implementations (first-order and S5) each position is stored as a tuple in the PROLOG database. While there are three conceptually distinct tuple types for representing quantifiers or modal operators, connectives, and atomic formulae, the same overall structure is used throughout.

The data fields in these tuples record such things as the polarity of the position, its descendents in the formula ordering, its label (the concrete syntax at that point in the formula tree), a pointer to its parent position, and a pointer to its next sibling in the formula tree. (Recall that the label of a position will be either a connective, a quantifier or modal operator, or an atomic formula.) These tuples are constructed and placed in the database as the formula is read in, and, with one exception described in the next section, are never modified thereafter.

Each tuple is assigned a unique numeric identifier as it is constructed.

All references to tuples by the theorem prover are made through the use of pointers. A pointer consists of a numeric identifer together with an index indicating the particular instance of the tuple being referenced.

This representation allows formula tree nodes to be accessed in a variety of ways other than by reference to their identifiers. For example, all occurrences of nodes containing a particular proposition or connective, or having a particular polarity, can be retrieved using the same sort of database lookup which retrieves the tuple associated with a particular identifier.

The assignment of numeric identifiers to tuples is done depth-first and left-to-right, so as to facilitate various frequently performed operations. For instance, if n and m are two position identifiers, and n' is the next sibling of n then m is dominated by n just in case (a) $n < m$ and (b) $m < n'$ (i.e., dominance can be determined using just two integer comparisons). For example, the signed modal formula $\langle \Box((A \wedge B) \wedge \Diamond C \Rightarrow \Diamond(A \wedge C)), 0\rangle$, represented by the tree shown in Figure 1, would be stored in the PROLOG database as shown in Figure 4.

Identifier	Syntax	Polarity	Type	Descendents	Next Sibling
(1,	$\Box$,	0,	π,	(2),	13)
(2,	$\Rightarrow$,	0,	α,	(3,9),	13)
(3,	$\wedge$,	1,	α,	(4,7),	9)
(4,	$\wedge$,	1,	α,	(5,6),	7)
(5,	A,	1,	$-$,	$-$,	6)
(6,	B,	1,	$-$,	$-$,	8)
(7,	$\Diamond$,	1,	π,	(8),	10)
(8,	C,	1,	$-$,	$-$,	11)
(9,	$\Diamond$,	0,	ν,	(10),	13)
(10,	$\wedge$,	0,	α,	(11,12),	13)
(11,	A,	0,	$-$,	$-$,	12)
(12,	C,	0,	$-$,	$-$,	13)

Figure 4: Example modal database.

4.1.2   Multiplicity and prefixes.

The modal multiplicity records the number of distinct instances of subformulae dominated by a ν type modal operator allowed within the (partial) proof at any given point in time. Positions are always considered indexed as described in section 3.1. The modal multiplicity is represented as a table with one entry for each ν_0 position in the formula tree. Indices are therefore stored as lists of positive integers. For S5, prefixes are simply positions which represent modal operators.

4.1.3   Proof Tree Representation

The proof state within a connection calculus is represented by a record of the connections made so far, the substitution constructed and the paths through the formula still to checked for the current multiplicity. OR choices arise when there is more than one possible connection that makes a given path complementary (see Section 4.2.1). The OR choices are maintained as a tree.

In this implementation, each node of the tree is stored in the database as it is constructed and identified by a string of bits. The identifier for a node N is constructed according to the following rules:

1. The identifier of the root node is 1.

2. If N is the right descendent of a node M and the identifier of M is B, the identifier of N is $1B$.

3. If N is the left descendent of a node M (or only descendent, if M is a unary node), and the identifier of M is B, the identifier of N is $0B$.

When these rules have been followed, the node whose identifer is I_1 is an ancestor of the node whose identifier is I_2 iff I_1 is a right-adjusted substring of I_2. This test can be done directly on the identifiers themselves, without requiring any database lookups or tree traversals. However, this test is more complicated, and hence slower, than the corresponding integer comparison test used to determine ancestry of formula tree nodes.

Storage of node identifiers is also more complicated for the dynamic proof trees than for the static formula trees. A node at depth D requires a bit-string D bits long to identify it. For trees of reasonable depth, D exceeds the length of the standard PROLOG integer, and so the bit string must be split across several such integers. In practise, identifiers and pointers are recorded using a three-place data structure, "code(X,Y,Z)", in which "Z" is a list of 16-bit integers, and "X" and "Y" identify the word and bit position containing the bit most recently added to the identifer.

Considerable savings in space could have been obtained if the search strategy was confined to depth-first, left-to-right through the space of possible connections. (Such a procedure resembles the search strategy of standard PROLOG interpreters.) However the original program was designed to investigate aspects of heuristic search, so a more flexible representation of partial proofs was required.

4.1.4   Connection graphs.

The last static data structure of interest is the *connection graph*. Since only a single copy of the formula tree is kept, the positions in which a particular atomic formula may be found with a particular polarity can be tabulated as the formulae are being input. This table is called a "connection graph", and by referring to it while constructing proofs the theorem prover never needs to search the formula tree to find complementary propositions. The connection graph is propositional but could be extended to include unification information.

Note that the use of a connection graph is distinct from the use of connection graph *resolution* [Kow75] in which alterations to this data structure are a part of the inference mechanism. The connection graph used here is not manipulated or changed in any way once the proof process is initiated.

Remark. However the two systems are in some interesting sense complementary. Connection graph resolution could be used to make permanent alterations in the formula tree so as to "compile" certain inferences and cut down on search within a run of the connection method. To our knowledge, this potential has not been exploited. ∎

4.2   Dynamic features.

4.2.1   Path checking and goal representation.

The connection theorem (in the case of S5, Theorem 3.4.1) underlying a given connection calculus expresses the validity of a formula in terms of a condition on the set of atomic paths through the formula. The main component of that condition is that a set of connections can be found such that every atomic path contains at least one connection from that set. Checking a formula for validity therefore reduces in part to a process of *path checking*. Indeed, many resolution based proof procedures can be interpreted in this way and a comparison made between them [Bib82b].

This path checking process can best be expressed using the *matrix* representation of the formula introduced above in which the components of a β-type formula are displayed one above the other in the plane whilst the components of an α-type formula are displayed side by side. The algorithm we use to check the atomic paths through a formula is based on that presented in [Bib82a]. A partial proof is represented by a set of *goals*. Each goal represents a set of atomic paths which have not yet been fully examined, but which share the same initial segment. The initial segment itself is divided into two parts, the *active path* and the *expansion*. The active path is a set of atomic formulae amongst which no connections exist. When choosing between possible connections, those involving the atomic formulae referenced in the active path are preferred over all others. This is equivalent to the unit preference heuristic [CL73]. The expansion is a set of references to partially examined subformulae. These references are generated by the reduction of α-type nodes, and correspond to dynamically created hypotheses which may contain extra problem-specific information in the form of variable instantiations not present in the original axioms. When choosing between possible connections, those involving the subformulae referenced in the expansion are preferred over those involving uninstantiated axioms. This is similar to the "set-of-support" heuristic used in resolution based systems.

4.2.2   Unification.

The reduction ordering is represented by a graph whose nodes are prefixes and whose arcs indicate precedence and/or equivalence of nodes. When a connection is formed, and two prefixes are unified, one new arc is added to this graph to show the effect on reduction order of the unification, and zero or more arcs added to show any new precedence information introduced by formula ordering. (Recall that the reduction ordering ◁ is the transitive closure of the union of the substitution relation and the formula ordering.) If at any point this graph becomes cyclic as a result of the introduction of arcs and/or nodes to represent a particular unification, the connection inducing that unification is inadmissible.

The unification of prefixes in S5 is equivalent to unification in a subset of first-order logic containing only constants and variables, and not functions or terms. Consequently, the representation used for prefixes, and the way in which they are unified, can be simpler than the corresponding mechanisms within the implementation for first-order logic. For example, a substitution table must be maintained when performing unification in first-order logic to allow the two unifications $x \leftrightarrow f(a, b)$ and $x \leftrightarrow f(b, a)$ to be distinguished. No such table is needed when unifying prefixes in S5 on the other hand, because all such unification is one-to-one. Similarly, unification of S5 prefixes is non-recursive, since there is never any need to "unpack" a term so as to unify its subterms.

4.2.3   Reporting proof search.

In [Wal86] it was shown how a connection-based search for first-order logic could be interpreted as constructing a generalised form of sequent (or tableau) proof. The derivations are "generalised" because the choice of prefix at the reduction of a ν type position is delayed until a later date via the process of unification. The derivations are similar in spirit to those constructed using Jackson and Reichgelt's sequent systems [JR87], or modal versions of Reeves' tableau system [Ree87].

In the implementation, this correspondence was exploited by reporting the search for connections as a sequence of applications of Kanger's sequent rules.

Remark. It is important to realize that the search is conducted at the level of connections. The use of sequent rules for proof output is simply a matter of "pretty-printing" to provide informative output. This should be contrasted with sequent based systems that search a much larger space, full of redundancies, in order to obtain such "natural" reporting facilities [BT75].

■

While this is a powerful result, many problems remain. We shall mention two.

Firstly, the naive translation of connections into sequent derivations (partial proof trees) does not produce an intuitively satisfying sequent proof (even though this proof is correct). The problem lies in the fact that although each individual connection can be reported as an intelligible derivation, the juxtaposition of derivations corresponding to successive connections does not necessarily result in any recognisable pattern.

Attempts were made to change the order in which reduction steps were displayed so as to make the resulting output more comprehensible, including printing some sequences bottom-up and others top-down, but no simple solution was found. Some *ad hoc* mechanisms incorporated into the theorem prover were useful. For example, when a position representing an implication is reduced, the theorem prover will print either

$$\text{Rule} : - > \text{C} \quad \text{or} \quad \text{Rule} : \text{A} - >$$

depending on whether the consequent or antecedent branch of the formula tree is being pursued.

Work by Andrews and his colleagues [And80,Mil84] may prove useful in this context even though these authors are concerned with the presentation of a *complete* proof expressed in terms of connections as a natural deduction or sequent calculus proof. Here we are concerned with the reporting the proof search directly.

The second problem concerns S 5 and modal logics in general. For these logics even well-formed sequent proofs are not intelligible since propositions may be true in one context and false in another; sequent proofs are not appropriate for the display of these contexts. A display facility based on the box diagrams of [HC68] is one possible approach.

5   Related work.

A number of authors have developed computationally oriented proof systems for S 5. Some are based on clausal resolution (*eg.*, [Far83]) since S 5 does admit a modal clausal form. Bibel's comparison of the classical connection calculus with classical clausal techniques in [Bib82b] suffice to demonstrate the benefits of our approach.

Abadi and Manna [AM86] develop a system based on non-clausal resolution. Their system suffers from various combinatorial problems due to the fact that the modal operators are manipulated by special deduction rules. The application of resolution is severely restricted. Our use of prefixes and unification removes the need for the special rules and liberates the use of the basic resolution operation (making a connection).

Konolige system [Kon86] involves the construction of multiple tableaux. Resolution is restricted to operate within each tableau. Again the combinatorics of his system are not ideal; indeed the construction of an auxiliary tableau is only justified in retrospect introducing many sources of redundancy. Our use of prefixes eliminates the need to consider multiple tableaux.

6   Conclusions.

We have presented the theory and some details of an implementation of a computationally efficient proof system for S 5 modal logic. The system is a variant on Bibel's connection calculus for first-order logic. The implementation was developed in a straightforward manner from a previous implementation of the connection calculus for first-order logic.

The basic techniques, summarised at the beginning of Section 3.1 were:

> The search should be confined in the first instance to selecting instances of the basic sequent schema of the appropriate sequent calculus. A process of unification together with a reduction ordering should be used to ensure that a correct sequent proof could be constructed at every stage without overcommitting the system to choices of particular values for the "variable" constructs. (In the case of first order systems these variable constructs are universally quantified variables. In the modal case the constructs are prefixes denoting possible worlds.)

It was suggested in [Wal86] that the techniques applied here to what is in effect the simplest modal logic could be used to produce computationally efficient proof systems for more complicated logics. We have succeeded in constructing a generalised connection calculus for the modal logics K, K4, D, D4, T, S4 and S5, as well as both constant- and varying-domain versions of their first-order variants. The details can be found in [Wal87]. In the latter case we can deal with versions of the logics in which the so-called "monotonicity" condition holds between worlds and versions in which no such condition holds. The generalised connection calculus is specialised to a particular version of a particular logic by altering the admissibility conditions on the modal substitution. All other details (including the methods for proof search) remain the same.

Acknowledgements

This research was supported in part by SERC/Alvey grants GR/D/44874 and GR/D/44270, and a British Council Commonwealth scholarship to the second author. We are indebted to Richard O'Keefe who described the utility of the numbering method for formula trees adopted, and Jane Hesketh who helped develop the bit string representation of integers in PROLOG.

References

[AM86]   M. Abadi and Z. Manna. Modal theorem proving. In J.H. Siekmann, editor, *8th International Conference on Automated Deduction*, pages 172–189, July 1986. Lecture Notes in Computer Science, Volume 230, Springer Verlag.

[And80] P.B. Andrews. Transforming matings into natural deduction proofs. In W. Bibel and R. Kowalski, editors, *5th International Conference on Automated Deduction*, pages 281–292, 1980. Lecture Notes in Computer Science, Volume 87, Springer Verlag.

[And81] P.B. Andrews. Theorem-proving via general matings. *Journal of the Association for Computing Machinery*, 28(2):193–214, April 1981.

[Bib81] W. Bibel. On matrices with connections. *Journal of the Association for Computing Machinery*, 28(4):633–645, October 1981.

[Bib82a] W. Bibel. *Automated Theorem Proving*. Friedr. Vieweg & Sohn, Braunschweig, 1982.

[Bib82b] W. Bibel. A comparative study of several proof procedures. *Artificial Intelligence*, 18:269–293, 1982.

[BT75]   W.W. Bledsoe and M. Tyson. *The UT Interactive Prover*. Research Report ATP-17, Departments of Mathematics and Computer Sciences, University of Texas at Austin, May 1975.

[CL73]   C-L. Chang and R.C-T. Lee. *Symbolic logic and mechanical theorem proving*. Academic Press, 1973.

[Far83] L. Fariñas-del-Cerro. Temporal reasoning and termination of programs. In S. Amarel, editor, *8th International Joint Conference on Artificial Intelligence*, pages 926–929, 1983.

[Fit72] M.C. Fitting. Tableau methods of proof for modal logics. *Notre Dame Journal of Formal Logic*, XIII:237–247, 1972.

[Fit83] M.C. Fitting. *Proof methods for modal and intuitionistic logics*. Volume 169 of *Synthese library*, D. Reidel, Dordrecht, Holland, 1983.

[HC68]   G.E. Hughes and M.J. Cresswell. *An Introduction to Modal Logic*. Methuen, London, 1968.

[HM84]   J.Y. Halpern and Y. Moses. Knowledge and common knowledge in a distributed environment. In *3rd ACM Conference on the Principles of Distributed Computing*, pages 50–61, 1984.

[HM85]   J.Y. Halpern and Y. Moses. A guide to the modal logics of knowledge and belief:preliminary draft. In *9th International Joint Conference on Artificial Intelligence*, pages 479–490, 1985.

[JR87]   P. Jackson and H. Reichgelt. A general proof method for first-order modal logic. Submitted to IJCAI-87, 1987.

[Kan57]   S. Kanger. *Provability in logic*. Volume 1 of *Stockholm Studies in Philosophy*, Almqvist and Wiksell, Stockholm, 1957.

[Kon84]   K. Konolige. *A Deduction Model of Belief and its Logics*. PhD thesis, Stanford University, 1984.

[Kon86]   K. Konolige. Resolution and quantified epistemic logics. In J.H. Siekmann, editor, *8th International Conference on Automated Deduction*, pages 199–208, July 1986. Lecture Notes in Computer Science, Volume 230, Springer Verlag.

[Kow75]   R. Kowalski. A proof procedure using connection graphs. *Journal of the Association for Computing Machinery*, 22(4):572–595, 1975.

[Mil84]   D.A. Miller. Expansion tree proofs and their conversion to natural deduction proofs. In R.E. Shostak, editor, *7th International Conference on Automated Deduction*, pages 375–393, May 1984. Lecture Notes in Computer Science, Volume 170, Springer Verlag.

[Moo80]   R.C. Moore. *Reasoning about knowledge and action*. Technical Note 191, SRI International, Menlo Park, Ca., 1980.

[Pne77]   A. Pneuli. The temporal logic of programs. In *18th Annual Symposium on Foundations of Computer Science*, pages 46–57, 1977.

[Ree87]   S. Reeves. Semantic tableaux as a framework for automated theorem proving. In J. Hallam and C. Mellish, editors, *(This proceedings)*, Wiley & Sons, 1987.

[Smu68]   R.M. Smullyan. *First-Order Logic*. Volume 43 of *Ergebnisse der Mathematik*, Springer-Verlag, Berlin, 1968.

[Sti85]   C. Stirling. *Modal logics for communicating systems*. Technical Report CSR-193-85, Dept. of Computer Science, Edinburgh University, 1985.

[Wal83]   L.A. Wallen. *Towards the Provision of a Natural Mechanism for Expressing Domain-Specific Global Strategies in General Purpose Theorem-Provers*. Research Paper 202, Dept. of Artificial Intelligence, Edinburgh, September 1983.

[Wal86]   L.A. Wallen. Generating connection calculi from tableau- and sequent-based proof systems. In A.G. Cohn and J.R. Thomas, editors, *Artificial Intelligence and its Applications*, pages 35–50, John Wiley & Sons Ltd., 1986.

[Wal87]   L.A. Wallen. *Matrix proof methods for first-order modal logics*. Research paper , Dept. of Artificial Intelligence, Edinburgh, 1987. (To appear).

[Wil86]   G.V. Wilson. *Implementation of a connection method theorem-prover for S5 modal logic*. Master's thesis, Department of Artificial Intelligence, University of Edinburgh, 1986.

Grammars and Natural Language

A NATURAL LANGUAGE INTERFACE FOR EXPERT SYSTEMS:
SYSTEM ARCHITECTURE

David P. Frost

Biomedical Computing Unit
Imperial Cancer Research Fund
Lincoln's Inn Fields, London   WC2A 3PX

0. Abstract

Interactions with expert systems are typically system dominated.
Provision of a natural language interface provides one means of
exploring mixed initiative dialogue. This paper outlines the overall
architecture of a natural language interface for medical expert
systems. It argues that traditional expert system design cannot
support the range and detail of natural language input. Instead a
knowledge base management system (KBMS) is proposed, modelled on the
functionality of database management systems. A dialogue interface
mediates between the natural language front-end and the KBMS. The
modularity of this design is discussed and a prototype implementation
reported.

1. Introduction

The use of natural language interfaces with expert systems has been
advocated for a variety of reasons. User convenience is the most
obvious one and yet in the short term it is the least important.
Typed natural language is not the most convenient means of
communicating with a machine, and connected speech input is still
some way off, (Johnson 1985). However as the complexity of expert
systems increases, or the user wants to take more initiative, so
natural language may become necessary for adequate communication.
(Sparck Jones 1985). Natural language may also reduce the
constraints on expert system design imposed by limitations in
existing user-interfaces.

Most work in applied natural language processing has been conducted
on front-ends to database systems. The considerable progress already
made in this area is a reflection of the highly structured nature of
database transactions and the existance of a underlying relational
data model, (Hendrix 1982). Recent work on database front-ends has
focussed on issues of cooperativity and inference within the natural
language component.

Expert systems pose a whole new range of problems not encountered in
designing front-ends to databases. These include the maintenance of
a dialogue with the user, and the handling of questions about the
system's internal reasoning. In medicine, for example, dialogues

between doctors and specialists frequently begin with the
presentation of a clinical vignette, such as -

> The patient is a 34 year old man with a three week history
> of breathlessness.
> He complains of a productive cough which gets worse at
> night.
> In the past, he has been treated for asthma with
> salbutamol.
> On examination, he had widespread crepitations and a
> pleural rub was heard at the right base.

The processing of such text is clearly non-trivial and without some
paraphrasing is beyond the capability of most current natural
language systems. Nevertheless case summaries of this form provide
an indication of the types of linguistic issues that need to be
addressed, and the level of detail that the expert system's knowledge
representation has to support. The system also needs to handle a
range of queries. For example

> Does the patient have a cough?
> What are the features of acute bronchitis?
> Does he show signs of left ventricular failure?
> How did you conclude that he had a respiratory infection?

These questions range from simple retrieval of data or knowledge, to
requests involving object-level or meta-level inference.

There are still only a few reports of natural language interfaces to
expert systems (e.g. Carbonell 1983, Lubonski 1985), and it is too
early to assess these in detail or identify any common approach to
design. It has been predicted that expert systems will need much
closer coupling (or integration) with the natural language front-end
than has been necessary with databases.

The current project addresses the following problems -

[1] To what extent does the natural language interface have to be
     integrated with the expert system?

[2] How does this integration effect the design of the expert system
     itself?

The project is set within the wider context of research on the
application of AI in medicine. This demands a pragmatic approach to
design and implementation. There is still a need for working
prototype systems which are capable of addressing real clinical
situations. The development of such systems involves the integration
of techniques many of which have been previously reported only in
isolation.

This paper focusses on the overall design of a natural language

interface for consultative expert systems in medicine. It proposes a
modular system architecture and discusses some of the issues arising
from this modularity. The paper also reports on a prototype
implementation of the system.

2. Architecture

A major criterion in designing the overall architecture of the system
was that it should be modular. This helps to clarify the functions
of the various component parts of the system and to define their
inter-relationships clearly. Modularity is also desirable from a
computational point of view, in that it limits the effects of local
changes. This is particular important in prototype AI systems which
are necessarily in a state of continuous evolution.

One potential line of demarcation is between the front-end processing
concerned with natural language and the 'back-end' processing in the
expert system. If it were possible to maintain a clear distinction
between these two activities, then some straightforward interface
between them could be built.

In portable natural language front-ends to databases a clear
modularity is maintained between the front-end and the underlying
database. An interface is also discernable between purely linguistic
parts of the front-end and those parts concerned with generic
database query generation, (e.g. Ginsparg 1983). The query
generating algorithm incorporates a data model reflecting the
conceptual structure of the database.

In expert systems design, the distinction between a front-end and a
back-end is more blurred. Typically the process of query evaluation
is closely connected with the management of the user interaction.
The reasons behind this integrated design are discussed later in
section 4.1. Despite this integration, expert systems do contain
knowledge which is solely concerned with the specific domain of
expertise and which is independent of the details of the dialogue
with the user.

In an attempt to draw a distinction between manipulations of the
knowledge base and interactions with the user, an intermediate
module, called the 'dialogue interface', was introduced into the
architecture. This mediates between a linguistic front-end and
knowledge base management system (KBMS) at the back.

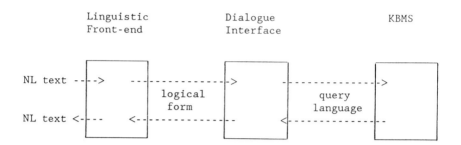

The inclusion of a dialogue interface brings the other two modules into clearer focus. The linguistic front-end is concerned solely with linguistic encoding, whilst the knowledge base system handles the representation and manipulation of domain knowledge.

Communications between the modules is strictly controlled. The data flow between the front-end and the dialogue interface is by means of logical forms which are independent of surface linguistic considerations. The back-end KBMS is addressed using a specific knowledge manipulation or query language.

The system is driven by input from the user. This is processed by the front-end to provide an input to the dialogue interface. The top-level goal of the dialogue interface is to generate an appropriate response. In doing so it may make any number of calls to the KBMS with updates or queries about either dialogue management or the specific domain of expertise. The front-end finally 'encodes' the response in natural language. The role of the dialogue interface is discussed in more detail in Section 5.2.

3. Linguistic Front-end

The design of the linguistic front-end of the system uses established techniques within computational linguistics. In keeping with the overall pragmatic approach towards design and implementation, no attempt is made to provide unrestricted coverage of English nor to handle ill-formed input. Where a particular linguistic feature is not provided, adequate alternatives are available requiring only minimal paraphrase, e.g. expanding out sentences containing conjunctions. The overall aim of the natural language front-end is to support an experimental dialogue between user and system.

3.1. Natural language interpretation

Input sentences are parsed on the basis of the syntactic categories of the constituent words using a chart parser. The grammar is based on generalised phrase structure grammar, (Gazdar 1984), and includes slash features, sub-categorisation and meta-rules. The chart uses a combination of bottom-up and top-down control, (Thompson 1981). It is implemented in Prolog and exploits Prolog's built-in unification, (Frost 1986).

A compositional approach is adopted for the semantic interpretation of the input text. The parser builds an expression in lambda calculus which is subsequently reduced to produce a logical form, (Warren 1983). Individual word senses are not interpreted at this stage, although sub-categorisation indices are included where appropriate. For example, the verb 'complain' in a sentence 'X complains of Y' is represented by 'complain:31(X,Y)', where 31 refers to the particular verb phrase sub-categorisation.

The scope of noun phrases is represented using three-branched quantifiers, (Dahl 1979). The three arguments (or branches) are a variable, its range, and its scope, (see example below). Wide scope quantification is handled by storage and reordering, (Cooper 1983). In the present system, a simple heuristic is used to define a linear ordering based on both quantifier strength and surface order.

A typical logical form is

```
exists( X,
    true,
    exists( Y,
        isa(Y,rub:0) & mod(Y,pleural),
        definite( Z,
            isa(Z,base:0) & mod(Z,right),
            exists( E,
                isa(E,hear:2) & time(E,past) & mod(at,E,Z),
                event(Z,X,Y) ) ) ) )
```

    USER: a pleural rub was heard at the right base

The initial logical form generated by the parser can contain references to existing discourse entities. Co-referencing these variables is a recursive process at each stage of which a 'quantifier collar' - a variable and its range - is removed. The range of existentially quantified variables is translated into new discourse entities. These are added to an ordered set of discourse entities which is built up during the dialogue. The ordering is based on recency of reference.

Definite quantified variables are coreferenced by matching their range against the existing discourse entities. Pragmatic (inter-sentential) anaphora are resolved by reference to a focus model based on Sidner (1983). This represents the current focus, potential focus and a focus stack. It also represents the current temporal focus using an interval semantics approach, (Allen 1981).

The removal of the quantifier collars reduces the logical form to a simple conjunction of predicates, except when some universal quantification is present. However universal quantification seems mainly to be concerned with the statement of general facts found, for example, during knowledge acquisition. In problem-specific dialogues, little use is made of universal quantifiers.

Ambiguity in the input sentence can lead to more than one interpretation. When this occurs, the output of the front-end consists of a set of alternate logical forms.

3.2. Natural language generation

A distinction is maintained between a strategic and a tactical stage of natural language generation (McKeown 1985). The strategic stage, which is responsible for the selection and organisation of information, forms part of the dialogue interface module. The tactical stage is part of the linguistic front-end. It employs the same syntactic and semantic grammar rules as the parser although for generation these are represented as a definite clause grammar (DCG).

The surface form of noun phrases is optimised to generate the least complex non-ambiguous construction for each phrase. Personal pronouns are introduced by reference to the focus model and set of current discourse entities. There is no optimisation of structures above the level of noun phrases.

Natural language generation is important not only for producing replies to queries. Since all internal representations within the system consist of logical forms, they all should have a translation into natural language sentences.

4. Knowledge-based system

One of the major problems in designing a natural language interface for expert systems is the lack of an established expert system architecture. This reflects the absence of any clear theoretical underpinning to the field. The approach adopted in this project is to attempt to define a clear distinction between knowledge manipulation and dialogue management. In doing so many of the features of traditional expert system design are lost. The extent to which this imposes restrictions on the functionality of the overall system remains to be determined.

4.1. Expert system design

The distinction between expert systems and knowledge-based systems (in general) is very blurred but there seem to be two ways in which they differ.

[1]   Expert systems tend to address specialised domains of expertise rather than commonly held knowledge.

[2]   Expert systems tend to have a very limited style of interaction with the user, ie. dispensing specialist information, advice or

criticism about particular problems.

This specialisation of function influences both the choice of knowledge representations and the strategic aspects of the design. In the simpler expert systems design, questions for the user are generated from within the search algorithm whenever some required data items are not found. Even if the control knowledge is made more explicit in terms of strategic rules, the distinction between query evaluation and dialogue management can remain blurred. The system's control strategy is concerned with both optimising the query evaluation using heuristic knowledge and also generating an appropriate order of questions for the user using knowledge about dialogue structure. The search tree can come to represent a dialogue plan.

The architecture proposed here maintains a clear distinction between query evaluation and dialogue management. Query evaluation is handled by a knowledge base management system (KBMS). There is no facility for the KBMS to generate questions directly nor for special access to its internal execution path. Instead the KBMS provides options for returning sets of potential questions for the user. The dialogue interface selects which of these (if any) should be put to the user.

4.2. Knowledge representation

The marked specialisation of function, which characterises expert systems, makes it possible for expert systems to employ very specialised and abbreviated knowledge representations. These representations, although simple, are able to capture significant aspects of domain expertise, (Buchanan 1984). Query evaluation relies on pattern-directed inference in which symbols are manipulated without reference to any semantic relationships that might exist between the symbols.

Whilst this approach can provide some measure of expert advice-giving, it cannot support activities such as teaching or explanation generation, (Clancey 1983). These activities require more explicit representation of structural knowledge about the domain. This kind of knowledge can be quite adequately represented using semantic network and frame-based systems, (Brachman 1979). However the inheritance mechanisms these provide cannot represent all aspects of domain expertise. There is still a need to represent relations of causality or empirical association between complex situations.

Take for example -

a cough with a prolonged lowering sound is one indication of damage to the left recurrent laryngeal nerve

This could be represented by a specific production rule of the form 'if P then Q', where P refers to the state of 'a cough with a

prolonged lowering sound' and Q refers to 'one indication of damage
to the left recurrent laryngeal nerve'. But neither P nor Q are
simple atomic situations. If they are represented as such then a
complete match is required with the data. There is no possibility of
exploring the implications of a partial match.

Production rules capture the concept of inference but do not support
an adequate description of the domain. Frame-based systems have a
much greater expressive power but cannot in themselves adequately
represent 'heuristic' knowledge, except via procedural attachment. A
third important data model is that found in relational databases.
Relational operations such as JOIN are similar in many ways to
inference rules, (Kowalski 1984). Transitive closure over reflexive
relations introduces some features of attribute inheritance found in
frame-based systems. Furthermore, the relational data model has a
sound theoretical basis as does the deductive database model
developing within logic programming, (Gray 1984).

The knowledge represented within an expert system, whether as facts
or rules, consists of propositions (or sentences) about the domain.
Consequently it is possible to consider the knowledge as a collection
of logical forms. This is the same type of representation as that
generated by the linguistic front-end. The knowledge representations
employed in the current system are all equivalent to logical forms
and consequently they all can be translated into natural language
sentences. Exploration of the implications of this choice of
knowledge representation is currently the central focus of research
in the project.

4.3. Knowledge base management system

The knowledge base management system (KBMS) is a modular structure
which responds to queries from the dialogue interface. Its
functionality is modelled on a database management system (DBMS).
Only one input and output stream is provided. The KBMS supports both
object-level and meta-level queries.

Updates consist of either additions or deletions to the database.
Currently no checking of integrity constraints is performed. Query
expressions are logical forms of either a yes/no or WH type. Options
are provided to select different inference mechanisms and to specify
the format of the reply returned by the KBMS, e.g. tables or trees.
Trees represent a summarisation of a table of results and are
obtained by using structural knowledge about the domain. The system
can also return a trace from the rule interpreter in the form of a
tree structure. These structures fit into the natural language
generation schema handled by the dialogue interface.

Options on query evaluation can also be used to specify the choice of
logic. The current system can return truth values using a multi-
valued logic, based on 3-valued logics, (Turner 1984). A distinction
is made between closed classes of data, in which there is negation by

failure, and open classes in which the truth values - true, false and unknown - are recorded explicitly. With open classes, failure implies that the data item is askable. Using this multi-valued system, the KBMS can return a structure such as -

        true/   disease:respiratory:viral
        false/   pneumonia
        unknown/ coryza
        askable/ if (joint:pain) flu

This states that pneumonia is not present but that a viral respiratory disease is. It is not possible to confirm that this is a common cold (coryza) but it may be flu if joint pain are present. It is left to the dialogue interface to determine which of a wide range of possible replies to select.

Meta-queries are handled in the same way as object-level queries although the data structures returned and the range of options available are different. Packaging of these results into an acceptable 'explanation' is a function of the dialogue interface.

5. Dialogue interface

The main function of the dialogue interface is to coordinate the system's activities within two quite different domains: (1) the domain of the dialogue, and (2) the domain of (medical) expertise. These represent the form and the content of the system's response to the user.

The domain of the dialogue concerns the representation of structures such as turn-taking, question-answer pairs and nested questions. It also involves strategic aspects of natural language generation including knowledge about response schemata. It is independent of the domain of expertise.

The domain of expertise concerns data about the current problem, either asserted or inferred. It is independent of the user, the user's goals or current focus of the dialogue. Its time scale is that associated with the problem domain. Within this domain, the ordering of updates and queries is immaterial.

5.1. Dialogue structures

The structure of dialogues between clients and experts has been investigated in various contexts. One common feature of consultations with experts is the client's wish to volunteer the relevant details of their particular problem, (Kidd 1985). There also a wide range of different questions, both factual and inferential, that can be asked, (see examples in Section 1).

A crucial issue in the processing of this wide variety of input is to decide what counts as an adequate response. Schemata for generating

definitions, descriptions and comparisons have been proposed by
McKeown (1985). Techniques for the generation of explanations have
also been reported, (Weiner 1980). However, strategies for
generating an appropriate 'form' to the response are needed for all
user inputs. Typically in expert systems, these strategies are built
into the architecture of the system and do not exist as explicit
knowledge.

5.2. Role of the dialogue interface

The overall role of the dialogue interface is the generation of an
appropriate response to the user's input, in the context of the
current dialogue. There are three levels within this process:

[1]   What sort of reply is wanted.

      The first task of the dialogue interface is to generate the
      range of possible formats for responding to the user. This
      requires knowledge about dialogue and the strategic aspect of
      natural language generation. This knowledge is maintained
      within the KBMS, as is a representation of the current dialogue.
      The top level of queries submitted to the KBMS by the dialogue
      interface therefore concerns how to respond to the 'form' of the
      user's input. The set of response strategies that are returned
      consist of conditional statements, whose conditional part
      specifies what is needed in terms of results within the problem
      domain. These are translated into problem domain queries.

[2]   What is needed and how to get it.

      The second level within the dialogue interface is concerned with
      the control strategies for evaluating problem domain queries.
      These strategies take the form of meta-rules which expand a top
      goal into an agenda of tasks. These tasks may themselves invoke
      further strategies or they may consist of object-level goals.

[3]   What the result is.

      The third type of activity within the dialogue interface is the
      evaluation of object-level goals. These are handled directly by
      the KBMS, which can return a variety of different structures,
      e.g. sets of potential questions, depending on the options
      selected.

The overall control strategy of the dialogue interface coordinates
these three levels within the global search space. Failure at each
level leads to backtracking and re-evaluation of goals. Use is made
of Prolog's internal control strategy.

One important feature of this design is that the activities of the
KBMS within the problem domain are subordinate to the needs of the
natural language generation schemata. The natural language

generation schemata form the major planning factor in dialogue.

6. Summary and Conclusions

This paper describes the overall architecture for a natural language interface to consultative expert systems. A major feature of this design is the division of the system into three modules - a linguistic front-end, a knowledge base management system and an intermediate dialogue interface.

Natural language front-ends to databases have a similar modularity of design. The most obvious feature of this is the modularity of the database itself which allows the development of portable front-ends. The front-ends usually consist of a linguistic analyser which generates some form of semantic representation. This is then processed by a translator stage which produces an appropriate expression in the query language of the database.

In marked contrast with this, traditional expert system design is insufficiently constrained to support a natural language front-end. The close connections between query evaluation and dialogue management mean there is no clear interface between front-end and back-end. Since expert system do not have (nor need) an explicit data model, there is little scope for developing an expert system query language. Such a query language is necessary in order to handling the range and detail of natural language questions.

The portability of the current system is very limited. KBMS design is only at an early stage and no formal query languages have been established. Until such query languages are defined, the only potentially portable aspect of an expert system is its knowledge base. The constraints that a clear conceptual data model make on the process of knowledge representation suggest that it should be possible to develop implementation-independent knowledge representations similar to the table format of relational databases.

The dialogue interface coordinates domain-specific reasoning with the management of the user dialogue. The introduction of this intermediate module allows research to be conducted on issues of dialogue management, e.g. cooperativity, independent of the linguistic front-end and the KBMS.

The construction of comprehensive natural language systems entails a considerable research effort in its own right, (Bates 1984). The system reported here is only a first approximation in which many issues remain unresolved. However the setting of this project within medical AI dictates a pragmatic approach to design and implementation. The work reported here aims to establish an overall design structure which can provide a framework for future research.

References

Allen, J.F. (1981) An interval-based representation of temporal knowledge. IJCAI 7 221-226

Bates, M. (1984) There still is gold in the database mine. COLING-84 184

Brachman, R.J. (1979) On the epistemological status of semantic networks. in Associative Networks (Ed. Findler) 3-50

Buchanan, B.G. and Shortliffe, E.H. (1984) The MYCIN experiments of the Stanford Heuristic Programming Project. Addison-Wesley

Carbonell, J.G., Boggs, W.M., Mauldin, M.L. and Anick, P.G. (1983) XCALIBUR Project report 1: First steps towards an integrated natural language interface. Tech Report, Carnegie-Mellon CMU-CS-83-143

Clancey, W.J. (1983) The epistemology of rule based expert systems a framework for explanation. AI 20 [3] 215

Cooper, R. (1983) Quantification and syntactic theory. D.Reidel

Dahl, V. (1979) Quantification in a three-valued logic for NL question-answering systems. IJCAI 6 182-187

Frost, D.P. (1986) A practical Prolog chart parser for GPSG. Tech Report, Imperial Cancer Research Fund, London.

Gazdar, G.J.M., Klein, E.H.,Pullum, G.K. and Sag, I. (1985) Generalised phrase structure grammar. Blackwell

Ginsparg, J.M. (1983) A robust portable natural language database interface. Proc. Conf. Applied Natural Language Processing 25

Gray, P. (1984) Logic, algebra and databases. Ellis Horwood

Hendrix, G.G. (1982) Natural language interface. Am J Comput Ling 8 [2] 56-61

Johnson, T. (1985) Natural language computing: the commercial applications. Ovum

Kidd, A.L. (1985) What do users ask ? - some thoughts on diagnostic advice. Proc. Expert Systems-85 9

Kowalski, R.A. (1984) Inference in expert systems. Alvey IKBS mailshot

Lubonski, P. (1985) Natural language interface for a polish railway expert system. Natural Language Understanding and Logic Programming (Ed. Dahl) 21

McKeown, K.R. (1985) Text generation: using discourse strategies and focus constraints to generate natural language text. Camb Univ Press

Sidner, C.L. (1983) Focusing in the comprehension of definite anaphora. Computational Models of Discourse (Ed Brady and Berwick) 267

Sparck Jones, K. (1985) Natural language interfaces for expert systems. Proc. Expert Systems-84 85-94

Thompson, H. (1981) Chart parsing and rule schemata in GPSG. Tech Report Dept AI Edinburgh RP-165

Turner, R. (1984) Logics for artificial intelligence. Ellis Horwood

Warren, D.S. (1983) Using lambda-calculus to represent meanings in logic grammars. Proc Conf Assoc Comput Ling 21 51

Weiner, J.L. (1980) BLAH, a system which explains its reasoning. AI 15 19-48

LEGAS
Inductive Learning of Grammatical Structures

Roman M. Jansen-Winkeln
Universität des Saarlandes
Federal Republic of Germany

Abstract

This paper describes the theoretical approach to acquiring simple grammar rules from examples and its realisation in a program called LEGAS (LEarning GrAmmatical Structures). The resulting rule set is developed incrementally through gradual generalisation and specialisation. After each learning cycle the rule set achieves a status called "*complete*" and "*consistent*" which guides the flow of control.

1 Introduction

Throughout recent years *Inductive Learning* has again become a focal field of artificial intelligence research. Both in theoretical considerations and in practical applications remarkable results have been achieved.

LEGAS takes up essential developments in the field of concept and rule learning. It is implemented in PROLOG. LEGAS induces grammar rules from examples. Strings are used as training instances. The examples are continuous in that they are neither described by a fixed number of attributes [HR84] nor by constructs of the desired target formalism [Emd84].

LEGAS combines mainly two typical learning tasks. First, it maps training instances from a given formalisms onto rules given in another formalisms – in this case "character strings" onto "transduction grammar rules" – and secondly it manipulates expressions of the target formalism, i.e. generalisation, specialisation and ordering of transduction grammar rules.

If we want to solve the first part of the task – rule hypotheses being generated and verified with examples – we have to adjust the techniques of concept learning, for instance "*specific conjunctive generalisation*", to the requirements of rule acquisition. Realising "*generate-and-test* cycles" is assumed by the evaluation strategy of PROLOG to be the best possible way.

The second part of the task – transduction grammar rules are analysed and manipulated – takes account of methods which A. Bundy et al. [BSP85] and R. S. Michalski [Mic84] describe in their analytic comparison of rule- and concept-learning systems. These methods are modified for application in LEGAS. Especially the evaluation strategy for transduction grammar rules has to be considered when applying these rules. LEGAS has a special data structure which, in the case of a sequential depth-first strategy, shows the procedural path to the formal consistency of a rule set.

PROLOG supports this procedure because of its ability to make meta-level inference, i.e. the ability to generate and manipulate clauses as well as to assess their effects on the clauses in the PROLOG database.

LEGAS has been used with great success for the acquisition of rules for the analysis of *compound nouns, verb inflection* and *plural formation* in German and for the *addition* of *dual numbers*. The learnable language class is the class of transducers [HU79], which, however, by using special heuristics can be extended to the class of recursively nested transducers. In order to demonstrate LEGAS in an understandable way, I have selected English noun-plural examples for this paper. More complex examples can be found in [Jan86].

The next chapter introduces definitions of the terms "consistent" and "complete", and a broad specification of the program algorithm is presented. The second half of the chapter gives a detailed description of the representation of examples and rules. After this, a procedure is presented for generating rule hypotheses and classifying the power of the learnable grammatical structures. The following chapter describes how inconsistencies within a rule set can be located and removed by the analysis of *"near misses"*. The last section deals with an attempt to find – by comparing this approach with similar approaches – essential features for the realisation of these systems.

2   Learning Algorithm for Transducers

LEGAS – a rule acquisition system – induces rules from examples. The process of learning is determined by three factors: examples, rules and the rule interpreter.

- **Examples:** Pairs of character sequences consisting of an input and an output form. The examples manifest the effect of the rules to be learnt.

- **Rules:** Functional interdependencies for the manipulation of strings. They can be shown explicitly and evaluated.

- **Interpretation:** Determines the procedural evaluation of rules when applied to strings. The interpretation is responsible for strategies of selecting rules and solving conflicts between them. An interpreter will be described which selects rules on the basis of an underlying partial order.

Summarising these factors we can say that they describe the State of the Learning Process (SLP) and control the incremental acquisition of a rule set.

2.1   The Learning Process

SLPs contain – apart from rules necessary for grammars – information on the chronological development of grammatical structures and a justification of the rules induced. In this respect the class of complete and consistent SLPs are of major importance. A State of the Learning Process is *complete*, if there is a deriving rule for every training instance. It is *consistent* if the rules can be ordered according to the relation *more specific than*. Within these SLPs, chronological data – consisting mainly of examples – are in line with the derivation scheme.

By using SLPs this inductive approach to learning may be described as directed transition from one complete and consistent SLP towards another SLP within the process state space.

The learning process starts with an empty example and rule set. A modification in the SLP is caused by adding new examples. If required, the rule set or its order are changed or extended.

This approach reflects the incremental nature of learning. The gradually increasing set of examples is constantly confronted with a set of plausible explanations. In addition, with this approach a first global specification of the inductive learning program LEGAS can be achieved.

Completness	Consistency	Action
–	?	generate new complete rules
+	?	acquire no more rules
+	–	modify existing rules or their order
+	+	rule set is complete and consistent

+ : Condition satisfied
– : Condition not satisfied
? : Condition not examined

According to this specification, rule acquisition can be subdivided into two phases:

- generation of rules completing an incomplete SLP. I call this phase *"acquisition of complete rules"*.

- modification of a rule set until the SLP is consistent. This is called *"generation of consistent rules"*.

2.2   Some Definitions

The following definitions describe the above stated facts formally.

Definition 1 *State of Learning Process (SLP):* A State of the Learning Process is described by a triple of the form $SLP := (D, R, SEQ)$. If A is an alphabet, then:

Example set D: The set of known training instances is realised by pairs of input and output forms.

$D \subset A^+ \times A^+$
$D = \{(a_i, z_i) | 1 \leq i, j \leq n, \text{ and } i \neq j \Rightarrow a_i \neq a_j\}$

Rule set R: The set of induced rules consists of partial functions on strings over alphabeth A.

$R \subset \wp(A^+ \times A^+)$
$R = \{r_j | 1 \leq j \leq k, \text{ and } r_j : A^+ \rightsquigarrow A^+\}$

Order relation SEQ: The sequence of rules and examples is described by the following triple:

$SEQ \subset R \times R \times A^+$

For all $r_1, r_2 \in R$ and $(a, z) \in D$ the set SEQ is defined as follows

$(r_1, r_2, a) \in SEQ$ if
- $r_1(a)$ and $r_2(a)$ have defined values;
- $r_1(a) = z_1$ and $r_2(a) = z_2$;
- $(a, z_1) \in D$ and $(a, z_2) \notin D$;

"Completeness" and "consistency" are required to assess States of Learning Processes.

Definition 2 *Completeness.* Let $PZ = (D, R, SEQ)$ be an SLP. PZ is called *complete* if

$$\forall(a_i, z_i) \in D, \exists r_j \in R \text{ with } r_j(a_i) = z_i.$$

Definition 3 *Consistency.* Let $PZ = (D, R, SEQ)$ be an SLP. PZ is called *consistent* if

$$\forall r_1, r_2 \in R : (r_1, r_2, a_i) \in SEQ \implies (r_2, r_1, a_j) \notin SEQ)$$

An input form $a \in A^*$ can be mapped onto an output form with the help of the corresponding derivation rule. In doing so the SLP selects the fitting *"derivation rule"* $r_j \in R$.

Definition 4 *Derivation Rule.* Let $PZ = (D, R, SEQ)$ be an SLP and $a \in A^+$ an arbitrary input.

$$RC(a) = \{r_j \in R | r_j(a) \text{ has a defined value}\}$$

If $RC(a) \neq \{\}$ then the *derivation rule* r_i for a will be found as follows:

- $r_i \in RC(a)$ and
- $\forall r_j \in RC(a) \setminus \{r_i\}$ holds: $\not\exists(r_j, r_i, x) \in SEQ$

Lemma 1 *All examples are derivable in an SLP if and only if the SLP is complete and consistent.*
A formal proof for Lemma 1 can be found in [Jan86].

2.3   Representation of the Learning Process

LEGAS is implemented in PROLOG. Thus, stylistic means typical of PROLOG are used for the representation of SLPs.

The rule set as well as the examples at the beginning of the induction are represented in a grammar formalism, called *"Transduction Grammar"* (TdG). The name was chosen since a transducer exists for each grammar – which may be recursively nested and augmented – that accepts and generates the same languages. The syntax of the TdG-production closely follows the notation of Definite Clause Grammars [PW80].

Productions consist of:

- one nonterminal symbol provided with an unambiguous rule index

- left and right sides of the production

Lefthand and righthand sides of a production consist of sequences of terminal and nonterminal symbols. Order and kind of nonterminal symbols are identical both on the right and left side of the rule whereas terminal symbols may show differences. Terminal and nonterminal symbols are presented as defined in Definite Clause Grammars.

Examples:

- Mapping of any chosen string onto itself:

  ```
  str/1 :  [] --> [] .
  str/2 :  [A], str --> [A], str.
  ```

  The first rule says that the empty string is mapped onto itself. In str/2 any terminal symbol at the beginning of an input form is matched by the variable A and, without being changed, placed at the beginning of the output form. The predicate str is applied recursively to the rest of the input form.

- Replacement of the character "y" by "ies" at the end of an arbitrary string.

  ```
  plural/i :  str, [y] --> str, [ies].
  ```

  The rule plural/i fails if the input form does not end with a "y".

TdG-rules are – like many other PROLOG predicates – bidirectionally applicable. They can be used to map input forms onto output forms and vice versa as well as to verify whether a given pair of input and output forms is derivable with a Transduction Grammar.

Each TdG-rule in PROLOG is provided – in contrast with many typical unification grammars – with an unambiguous index i. By adding this index to a nonterminal, the non-deterministic rule selection of the Prolog interpreter becomes invalid and the rule with the index i is applied.

The user can access the TdG-rules via a window-oriented interface. It can be used as a monitor when training instances are recorded and for the control of the rules induced (Figure 1). Furthermore, this interface provides access to the internal representation of TdG-rules in PROLOG (Figure 2).

3   Acquisition of Complete Rules

A complete TdG-rule is formed in three steps:

- Proceeding from the input forms of the pairs of examples, the rule domain is generated bottom-up.

- From the domain, hypothetical ranges can then be generated

- The resulting rule hypothesis is verified top-down against the output forms of the example pairs.

If one of these steps fails, alternative solutions are searched for in the preceding step (backtracking).

The algorithm terminates as soon as a rule – according to the above system – can be found for an example set or backtracking is carried out beyond the first step. In the latter case the example set is divided and the procedure is repeated for each sub-set.

The basic idea for the generation of rule hypotheses is the search for shared, significant substrings among all input and output forms. Following Michalski [Mic84], the act of generalisation is denoted by the operator "|<" When inducing rules from examples, the following procedure is applied:

Roman M. Jansen-Winkeln

```
------------------------------------------------------------------
INPUT FORM            LEGAS                   OUTPUT FORM
------------------------------------------------------------------


2:diplomarbeit            << ?? >>              diplom-arbeit
1:diplompruefung          << ?? >>              diplom-pruefung
RULE-HYPOTHESE:
  cmp/cmp1 : [d i p l o m],str --> [d i p l o m -],str

2:eingabe                 << ?? >>              ein-gabe
1:einbaum                 << ?? >>              ein-baum
RULE-HYPOTHESE:
  cmp/cmp2 : [e i n],str --> [e i n -],str

2:diplompruefungseindruck          diplom-pruefung-ein-druck
                      diplom-pruefungseindruck
1:diplomarbeitseinband             diplom-arbeit-ein-band
                      diplom-arbeitseinband
RULE-HYPOTHESE:
  cmp/cmp3 : cmp/cmp1,[s],cmp/cmp2 --> cmp/cmp1,[-],cmp/cmp2
------------------------------------------------------------------
LERNZIEL:             AKTION:                 TR:  TE:
compound-analysis.    ANALYSE                 2.   0.
------------------------------------------------------------------
```

Figure 1: Example-Recording and Rule-Tracing Window

- Significant substrings of the input forms are mapped onto significant substrings of the output forms.

- Insignificant substrings remain unchanged during mapping.

Example: First, all training instances of an example set

$$D = \{(lady, ladies), (body, bodies), (copy, copies)\}$$

are translated into basic rules, each of them describing the mapping of exactly one training instance. Then the following generalisation may be carried out:

$$\underbrace{[\text{lady}] -- > [\text{ladies}]} \quad \underbrace{[\text{body}] -- > [\text{bodies}]} \quad \underbrace{[\text{copy}] -- > [\text{copies}]}$$

$$|< \qquad\qquad |<$$
$$\underbrace{\text{str}, [\text{dy}] -- > \text{str}, [\text{dies}]} \qquad |<$$
$$|<$$
$$\text{str}, [\text{y}] -- > \text{str}, [\text{ies}]$$

```
cmp(_80,_81,_82,_83,cmp3) :-
    cmp(_80,_91,_82,_92,cmp1),
    _91=s>>_93,
    _92<<-=_94,
    cmp(_93,_81,_94,_83,cmp2).
cmp(_80,_81,_82,_83,cmp2) :-
    _80=e>>i>>n>>_93,
    _82<<e<<i<<n<<-=_94,
    str(_93,_81,_94,_83,_97).
cmp(_80,_81,_82,_83,cmp1) :-
    _80=d>>i>>p>>l>>o>>m>>_93,
    _82<<d<<i<<p<<l<<o<<m<<-=_94,
    str(_93,_81,_94,_83,_97).
```

Figure 2: Internal Representation of TdG-Rules

3.1 Specific Conjunctive Generalisation

The rule hypothesis in the example described above was unambigous. If, however, there is a rule for a set of training instances, then the induced rule hypothesis need not be the only one.

The set $VS_D \subset TdG$ of all possible rule hypotheses for an example set D is called the *"version space"*, elements of VS_D are called *"specific conjunctive generalisations"* (SCG). The rules of a version space can be partially ordered using the relation "more specific than". Here the minimal and maximal elements are of major interest.

Definition 5 *MaxSCG*. Rule r is a maximally specific conjunctive generalisation (Max SCG) if no additional terminal symbol can be integrated into any significant substring of r.

Definition 6 *MinSCG*. Rule r is a *mininally specific conjunctive generalisation (Min-SCG)* if no terminal symbol in r can be replaced by the nonterminal "str".

For the example set

$$D = \{(lady, ladies), (body, bodies)\}$$

there are, among others, the following rules in VS_D.

```
MaxSCG : str, [dy]--> str, [dies]
MinSCG : str, [y] --> str, [ies]
```

3.2 Acquisition of SCGs

Due to the distinction between MinSCG and MaxSCG, the basic system for the acquisition of complete rules has to be extended. The induction algorithm as described is modified in a way that the result is the MaxSCG of the version space of a given example set. The MinSCG corresponding to any MaxSCG can be generated by dropping terminal symbols.

The MaxSCG of an example set is accepted as a rule in the rule set of an SLP. This approach takes account of a "careful" gradual learning strategy which tries to avoid over-generalisation already at the point of rule formation. The transformations of terminal symbols to be found within a MinSCG allow a classification of rules according to various criteria. Rules with the same tranformation, i.e. the same MinSCG, are potential candidates for a subsumption to one rule.

If a rule r is found whose MinSCG coheres with at least part of the MinSCG of a rule s then r is a potential candidate for recursive application as nonterminal in s. In doing this, use is made of the selective, index-driven application of TdG-rules. (Figure 1 shows an application of this feature during compound-analysis.) Storing the MinSCG and its transformations by inverted indices make it possible to optimize rules efficiently within one SLP and to allow an increased indepedence from the induced rules of the presentation sequence of associated examples. Furthermore, rule sets already fully acquired for a learning objective X can be integrated into a new rule set for a learning objective Y. The advantage of this integration is that no additional consistency checks on rule set X need be carried out.

4 Generation of Consistent Rules

In the second phase of rule acquisition all rules generated by the examples and their substrings are to be integrated into an existing rule base.

In the first phase, rules were merely considered when needed in order to further generalise a rule to be acquired. When rules are inserted however, all rules of an SLP and their ordering must be taken into account.

Rules are to be inserted into the rule set in such a way that all examples can be derived correctly according to our broad specification. For this – according to Lemma 1 – the SLP must be consistent.

One inherent feature of rule sets is that several rules may be applied to certain input forms. The set of all applicable rules for any input form a is called "*conflict set* of a". This is why every interpreter uses *conflict solving strategies* to select the rule to be applied from a conflict set.

In order to integrate a rule consistently into a rule set, it has to be inserted in such a way that, when the respective SLP is used to derive training instances, there are either no rule conflicts or rule conflicts are limited to the case where the correct rule is selected. If a learning system is to fulfill this task, information is required on how the interpreter works when discovering and solving conflicts. The way this was realised in LEGAS will be presented in the following paragraph.

4.1 Inconsistent SLPs

The selection and application of the rules of a TdG correspond to a large extent to the selecting and processing strategy of standard PROLOG interpreters. The reason for this is that TdG-rules – similar to DCG-rules, for example – can be transformed directly into PROLOG clauses. Thus, the sequential order of TdG-rules is of direct relevance.

Regarding our learning system LEGAS, the relation SEQ is part of the SLP so that information as to the order of rules to be found in the PROLOG database can be evaluated. The elements of this relation contain information on which rules r_i are part of the conflict set of an input form a and which of the rules is to be applied to a.

By using SEQ two different causes of inconsistencies of the SLP can be stated:

- **Control Fault:** The order of the TdG-rules (determined by SEQ) and of the corresponding PROLOG clauses (in the PROLOG database) do not coincide.

- **Factual Fault:** For two rules r_i and r_j contradictory information in the relation SEQ can be found stating that both r_i is more specific than r_j as well as r_j is more specific than r_i.

4.2   Insertion of Rules into the Rule Set

In the first step, the newly acquired rules are added to the existing rule set. The PRO-LOG clauses – belonging to the new rules – are inserted arbitrarily into the PROLOG database. Normally, however, they are placed at the beginning of the database. In order to find out whether or not the extended rule set is still consistent, all training instances are processed and checked as to whether the rule used for derivation produces the correct results. If all examples were derived correctly, the rule set is complete and consistent according to Lemma 1. If not, the relation SEQ is established for the conflict set of all incorrectly derived examples. If, e.g., the examples belonging to rule r_i are derived incorrectly by rule r_j then all elements (r_i, r_j, e) and (r_j, r_i, e) of the relation SEQ are analysed. The input forms can be split into two subsets:

$$NM(r_i) = \{(a, z)|(r_j, r_i, a) \in SEQ \text{ and } (a, z) \in D\}$$
$$NM(r_j) = \{(a, z)|(r_i, r_j, a) \in SEQ \text{ and } (a, z) \in D\}$$

The abbreviation NM stands for *"near miss"*. This term was introduced by P.H.Winston [Win75] and designates those negative training instances for a concept which are very similar to the positive training instance. In contrast to Winston's system "ARCH", LEGAS does not predetermine near misses explicitly. LEGAS analyses positive training instances of one rule in order to determine whether or not they are suitable as near misses for other rules.

By carrying out case studies of the relation between near misses we can state causes for inconsistencies of a rule.

- $NM(r_i) \neq \{\}, NM(r_j) = \{\}$: In this case r_i can be applied to examples belonging to rule r_j. On the other hand input forms of examples to r_j cannot be processed by r_i. Consequently, r_j is more specific than r_i and the sequence of these rules must be reversed in the PROLOG database.

- $NM(r_i) \neq \{\}, NM(r_i) \neq \{\}$: In this case r_i can be applied to part of the examples belonging to r_j and vice versa. The rules r_i and/or r_j have to be specialised, i.e. their domains have to be constrained.

This spezialisation is done by

- calculating discriminating substrings within the example sets,

- searching predefined domain-specific descriptors such as *"vowel"* or *"consonant"*,

- proposing the acquisition of new domain-specific descriptors,

- splitting up an overly general rule into various more specific rules.

4.3 Example

The incremental creation of a rule set for the English plural could be done in the following way.

From an initial example set

$$D_1 = \{(lady, ladies), (copy, copies)\}$$

the complete rule, already introduced to the reader,

```
plural/1 :  str, [y] --> str, [ies].
```

is induced and accepted as the first rule in the rule set R.

Extending the example set by

$$D_2 = \{(dog, dogs), (cat, cats)\}$$

results in a new rule to be inserted consistently which, again, is placed at the beginning of the rule set

```
plural/2 :  str, [] --> str, [s].
plural/1 :  str, [y] --> str, [ies].
```

This sequence leads to inconsistencies of the SLP since, because of the sequential evaluation strategy, all input forms of the examples of the rule "plural/1" are processed by rule "plural/2".
Two sets of near misses are created

$$NM(\text{plural/2}) = \{(lady, ladies), (copy, copies)\}$$
$$NM(\text{plural/1}) = \{\}$$

Consequently, the transformations of the TdG-rules have to be reversed in the PRO-LOG database, leading to:

```
plural/1 :  str, [y] --> str, [ies].
plural/2 :  str, [] --> str, [s].
```

Another extension of the example set by the training instance

$$D_3 = \{(day, days)\}$$

does not require the generation of a new rule, since plural/2 derives this training instance correctly. Inspite of this, the new SLP becomes inconsistent, because plural/1 is applicable to "*day*", i.e. $(\text{plural/1}, \text{plural/2}, day) \in SEQ$.

A renewed exchange of rules would be of no avail. Since plural/1 is more specific than plural/2, plural/1 is constrained by a predefined domain-specific descriptor e.g. "consonant"

```
plural/1 :  str, consonant, [y] --> str, consonant, [ies]
```

If the descriptor "consonant" is neither known nor can be acquired, plural/1 is split up into

```
plural/3 :  str, [dy] --> str, [dies].
plural/4 :  str, [py] --> str, [pies].
```

Flow of Control:

	LEGAS	SERA	Oakey et al.
1st Phase	Collection of Training Instances	Phase of Evaluation	—
2nd Phase	Acquisition of Complete Rules	Setup Phase	Rule Generation
3rd Phase	Subsumption of Rules	Integration Module	Generalisation
4th Phase	Generation of Consistent Rules	Correction Phase	Rule Stabilisation

Methods and Data Structures:

	LEGAS	SERA	Oakey
Positive Training Instances as Support	yes	yes	no
Numerical Assessment	no	no	yes
Rule Representation	TdG-Rules	Horn Clauses	Domain-Specific Rules
Rule Evaluation	Horn Clauses	Lambda Expressions	Special Interpreter
Language Class	Regular Expressions	Primitive Recursive Functions	—
Criterion for Consistency	All Examples are Reproducible	—	All Examples are Reproducible

Figure 3: Analytic Comparison of LEGAS

5   Summary

5.1   LEGAS in Comparison with Other Systems

As far as its structure is concerned, LEGAS is in some respects similar to other inductive rule acquisition systems. If, for example, LEGAS is compared with the program for the acquisition of pronunciation rules by S. Oakey and R.C. Cawthorn [OC81] or with SERA[Bol84], a system for learning recursive algorithms, then in all systems, a first phase can be found for the generalisation of rules with the help of examples, a second for the improvement of rules and a third for the specialisation, if required.

The target formalism of each of these programs (e.g. TdG-rules in LEGAS) depends on the domain being learnt and describes its mappings in a domain-specific way. Their representation may be interpreted directly (Oakey et al.) or – for efficient evaluation – may be translated into different formalisms (LEGAS and SERA).

Differences are reflected in the formalisation of the obtainable learning results. For instance, both in LEGAS and in SERA upper limits of the learnable language classes are stated (regular expressions or primitive recursive functions). Additionally, LEGAS provides a formal definition of the criterion for consistency. The table in figure 3 presents an overview of the most important points of comparison.

5.2 Technical Data

LEGAS was implemented in the form of a PROLOG program. The program was developed at the Universität des Saarlandes, Department of Computer Science, with the support of Nixdorf Computer AG, Paderborn. LEGAS consists of about 150 predicates and is realised in about 4000 lines of commented code. It is implemented on a DEC-VAX 11/780 in interpreted C-PROLOG-1.5.

Learning the most essential rules for the English plural as well as some exceptions requires about one-and-a-half-minutes of CPU-time and about 1/2 megabyte main memory.

The ideas presented here will be further developed within the framework of the German Special Collaborative Program on AI and Knowledge-Based Systems (SFB 314) of the German Science Foundation (DFG). They will be integrated into the knowledge acquisition tool which supports the establishment of a knowledge-based, natural-language front end XTRA (Expert Translator) for interfacing various expert systems.

References

[Bol84]   Toni Bollinger. *SERA – Ein System zum Erlernen rekursiver Algorithmen.* Master's thesis, Universität Bonn, 1984.

[BSP85]   A. Bundy, B. Silver, and D. Plummer. An analytic comparison of some rule-learning programs. *Artificial Intelligence*, (27):137–181, 1985.

[Emd84]   Werner Emde. *Inkrementelles Lernen mit heuristisch generierten Modellen.* KIT-Report 22, TU-Berlin, 1984.

[HR84]   C. Habel and C.-R. Rollinger. Lernen und Wissensakquisition. In C. Habel, Editor, *Künstliche Intelligenz; Frühjahrsschule Dassel/Solling, März 1984*, Springer Verlag, 1984.

[HU79]   J. E. Hopcroft and J. D. Ullman. *Introduction to Automata Theory and Computation.* Addison-Wesley, 1979.

[Jan86]   Roman M. Jansen-Winkeln. *LEGAS - Ein System zum induktiven Lernen grammatikalischer Strukturen.* Memo, Universität des Saarlandes, D-6600 Saarbrücken, 1986.

[Mic84]   R. S. Michalski. A theory and methodology of inductive learning. In R. S. Michalski, J. G. Carbonell, and T. M. Mitchell, editors, *Machine Learning – An Artificial Intelligence Approach*, Springer Verlag, 1984.

[OC81]   S. Oakey and R. C. Cawthorn. Inductive learning of prononciation rules by hypothesis testing and correction. *7th International Joint Conference on Artificial Intelligence*, 1981.

[PW80] Pereira and H. D. Warren. Definite clause grammar for language analysis
– a survey of the formalism and a comparison with augmented transition
networks. *Artificial Intelligence*, (13), 1980.

[Win75] P. H. Winston. *Learning Structural Descriptions from Examples*. McGraw-
Hill, 1975.

DIRECT INTERPRETATIONS OF THE GPSG FORMALISM

Roger Evans

Cognitive Studies Programme,
University of Sussex, Falmer, BN1 9QN, UK

INTRODUCTION

Generalised Phrase Structure Grammar (GPSG)[1] is one of the most important recent developments in formal syntax in recent years, especially from the viewpoint of the computational linguist. Not only is GPSG a well-specified, formal theory covering a wide range of syntactic phenomena in natural languages, but it is founded on a base (context-free grammars) with well-understood computational properties. In this paper I look at the kinds of computational GPSG systems that have been proposed, and suggest that to date they have all fallen into a single mold, exploiting only some of the analytic mileage that GPSG provides. As an example, I discuss Shieber's approach to the parsing of ID/LP grammars, and propose a somewhat different ID/LP parser which, I claim, exploits aspects of the ID/LP sub-formalism other systems miss. More generally I suggest that a more comprehensive exploitation of GPSG benefits both the computational and the theoretical linguist.

THE STRUCTURE OF GPSG

The GPSG formalism provides a specification language in which one can write grammars that describe natural language syntax. Unlike most standard linguistic formalisms, however, GPSG grammars do not characterise languages directly: instead they describe grammars in the context-free formalism and it is these which are used to characterise languages in the standard context-free fashion. In order to determine the language characterised by a given GPSG grammar, then, two operations are required. Firstly, the GPSG grammar must be expanded to the 'underlying' context-free grammar, and then the language characterised by *that* grammar determined (see (1)).

(1)   GPSG grammar → context-free grammar → language

The two basic reasons why this high-level representation language is more useful than the context-free formalism on which it is based are its compactness and its explanatory power. The GPSG representation of a grammar is in general more compact than its context-free counterpart, because it can encode (by virtue of the interpretation of the grammar specified by the formalism itself) a range of generalisations about context-free rules. A familiar example of this is the use of feature variables (or underspecified category specifications) in a rule, abbreviating many context-free rules each of which has a different combination of feature values in place of the variables. The explanatory power of the GPSG representation derives from the high level understanding of the components of the grammar and their interactions. For example, GPSG meta-rules are used to generate new ID rules[2] from old in a systematic fashion. The meta-rule in (2) captures the relationship between active and passive verb-phrases.

(2)   vp → ..., np   ==>   vp[pas] → ..., (pp[by])

Briefly, (2) states that for every *vp* rule with an *np* daughter that is actually in the grammar, there is also a corresponding passive *vp[pas]* rule implicitly in the grammar with the *np* removed and optionally a *pp[by]* inserted. The presence of this meta-rule in a GPSG grammar, coupled with an understanding of the role of meta-rules in GPSG, explains why in the underlying context-free grammar every active verb-phrase has a passive counterpart: at the context-free level itself this seems little more than a coincidence.

A COMPUTATIONAL VIEW

Consider now a computational implementation of GPSG in, say, a recogniser which accepts a GPSG grammar and a phrase and determines whether the phrase is grammatical according to the grammar. The most straightforward approach to designing such a system would be simply to mirror the theoretical structure described above, resulting in a system which first expanded the GPSG grammar to the context-free form, and then executed a conventional context-free recogniser algorithm. However, this approach has several drawbacks, both practical and theoretical.

At the practical level, the underlying context-free grammars associated with serious GPSG grammars are so big, that even the most efficient known context-free recogniser algorithms on the biggest, fastest machines are unlikely to produce acceptable real-time responses. This is a measure of the abbreviatory power of the GPSG formalism, as well as the complexity of natural languages. More theoretically, such a system would not exhibit characteristics of any great interest to anyone but a theoretical computer-scientist. For example, one could not expect this 'brute force' approach to provide any insight into or confirmation of psychological theories of how language structure relates to linguistic processing.

Because of these drawbacks, systems of this sort are rarely put forward as candidate designs for serious consideration. Instead, most researchers develop systems in which some portion of the GPSG formalism is interpreted *directly*, in other words, without first being expanded to the fully specified context-free rule representation. Thus virtually every GPSG parsing system deals with partially specified categories by means of on-the-fly feature instantiation which depends on the case in hand, rather than pre-compiled instantiation into all the possible combinations. Many systems go further in their directness: ProGram (Evans and Gazdar 1984, Evans 1985) handles the 'foot feature principle' (the automatic transmission of *slash* categories, etc.) directly, as does the parser discussed in (Pulman 1983), (Shieber 1984) and (Kilbury 1984) discuss algorithms for handling ID and LP components directly, (Evans 1982) describes a parser in which HFC is handled directly, etc. etc.

What is happening in all these systems is that the practical difficulties of the full context-free form are being overcome by retaining some of the abbreviatory power of the original GPSG form, and only expanding when there is sufficient information (from the input phrase) to constrain the number of possibilities. Thus not only is the abbreviatory power of GPSG one of its important theoretical properties, but it contributes also to the practicality of parsing algorithms.

What, then, about the explanatory power of GPSG? It seems reasonable to ask whether that might make a computational contribution in a similar fashion. Not only might further efficiency gains be made, but such systems might plausibly have something of interest to offer regarding the theoretical issues of linguistic processing mentioned above, an area barely touched by the systems exploiting just the abbreviatory potential. Indeed, it is unlikely that research into systems of this latter type could make substantial contributions to GPSG as a theory, since to a large extent it is the explanatory aspects of GPSG that motivate its structure.

However, it appears there are no such systems. The computational approaches to GPSG explored to date seem all to fall within the mold of those mentioned above. They take the context-free formalism as the basis of the criteira for grammaticality and augment it only with techniques (admittedly often very clever techniques) for delaying expansion of the GPSG abbreviations to the full context-free representation for as long as possible, so that non-determinism in the expansion process can be minimised. My principal aim in this paper is to highlight this point, and to motivate the case for the development of systems which do exploit GPSG's *explanatory* value, as well as its abbreviatory value.

AN EXAMPLE

To make the distinction more concrete, I shall look at the ways in which the ID/LP component of GPSG has been interpreted to date, and how the alternative view I am advocating differs from it.

Some theoretical background: ID/LP grammars   ID/LP grammars were first introduced in (Gazdar and Pullum 1981). The basic idea is that a context-free rule conveys two distinct types of information: information about what combination of daughter categories are dominated by what mother category, and information about the ordering of the daughter categories. So for example, a context-free rule such as (3) (a) expresses both the facts in (b) and (c).

(3)   a)   A → B C D
       b)   An A dominates a B, a C and a D.
       c)   The B precedes the C which in turn precedes the D.

An ID/LP grammar is one in which these two types of information are represented separately, so that generalisations about dominance and order can be made independently. Thus instead of the context-free rule (3) (a), we write (4).

(4)   a)   A → B,C,D
       b)   B < C
       c)   C < D
       d)   B < D

Rule (a) is an *Immediate Dominance* (ID) rule, expressing only the dominance relation. The daughter categories form a *set*, rather than a sequence. Notationally, the use of commas between the daughter categories distinguishes an ID rule from a context-free (CF) rule. (b), (c) and (d) are *Linear Precedence* (LP) rules, making statements about the ordering of the daughters ($B < C$ is read as B precedes C, that is, B is to the left of C in the input phrase). The three LP rules in (3) ensure that the daughter categories must come in the order specified in (2). [3]

Clearly any single CF rule can be put into this ID/LP form, consisting of an ID rule and a set of LP rules. More generally a set of ID rules together with a set of LP rules represents a set of context free rules. For example, the ID rules in (5) (a) together with the LP rules in (b) represent the context-free rules in (c).

(5)   a)   A → B,C,D
             A → B,C,E

      b)   B < D
           C < E

      c)   A → B C D   A → B C E
           A → C B D   A → C B E
           A → B D C   A → C E B

At this point a few definitions will make the following discussions clearer. An *ID/LP couple* is a set of ID rules together with a set of LP rules. Any ID/LP couple is equivalent to a (possibly empty) set of CF rules. For example, (5) (a) and (b) together form an ID/LP couple, equivalent to the CF rules in (c). A *proto-ID/LP grammar* is a set of ID/LP couples. Any proto-ID/LP grammar is equivalent to a context-free grammar, obtained by taking the union of the sets of context-free rules associated with each couple in the grammar. Furthermore any context-free grammar is equivalent to some proto-ID/LP grammar, since as (4) showed, each individual CF rule can be represented as an ID/LP couple and hence the grammar as a whole can be represented by the set of all such couples. There may also be other proto-ID/LP formulations of a given context-free grammar – the proto-ID/LP grammar in (5) (a) and (b) for example, does not take this form.

Finally an *ID/LP grammar* is a proto-ID/LP grammar consisting of just one ID/LP couple. Once again, associated with any ID/LP grammar is a context-free grammar ((5) is an example), but this time the converse is not true. There are context-free grammars which cannot be expressed as ID/LP grammars. Consider, for example, the two-rule context-free grammar (6).

(6)   A → B C D
        A → D C B

This grammar can be put into proto-ID/LP format (any context-free grammar can), resulting in the two ID/LP couples in (7).

(7)   <{A → B, C, D}   {B < C,   C < D,   B < D}>
       <{A → B, C, D}   {D < C,   C < B,   D < B}>

But no single ID/LP couple will capture exactly these two CF rules. Such a couple would need only one ID rule ($A → B, C, D$) but would have no LP rules, since every daughter follows every other in one or other of the CF rules. Thus the only candidate ID/LP couple would be (8) (a), and this is equivalent to the *six* CF rules in (b).

(8)   a)   <{A → B, C, D}   { }>

b)   A → B C D   A → C D B
      A → B D C   A → D B C
      A → C B D   A → D C B

The problem here is that using ID/LP grammars, there is no way to specify a grammar containing the two rules in (6) but not also containing the other four rules on (8) (b).

Hence the notion of ID/LP grammars identifies a strict subset of the context-free grammars [4]. Returning to the theme of abbreviation versus explanation, while a proto-ID/LP grammar may sometimes *abbreviate* a context-free grammar (employing fewer rules to capture the same information), it provides little *explanation* of its structure. On the other hand an ID/LP grammar both abbreviates and provides explanation: giving a reason, in the higher level terms of closure under legally ordered permutations of daughter categories, why one CF grammar is acceptable, but another is not, why the presence of certain CF rules (e.g. those in (6)) necessarily entails the presence of others (the other four in (8)). One of the assumptions made in the GPSG formalism is that GPSG grammars are ID/LP grammars, amounting to the claim that natural language grammars fall into this particular subset of context-free grammars, *and* providing some structural explanation of this fact.

Existing computational interpretations of ID/LP grammars   Since   the introduction of ID/LP format into GPSG, there have been several direct computational interpretations of it, for example (Shieber 1984, Kilbury 1984, Evans 1985, Barker 1986). For the purposes of this discussion, I shall fix attention on the earliest one, (Shieber 1984), but the main points of the argument hold equally for the others.

Shieber's algorithm is based on Earley's parser for context-free grammars (Earley 1970). The fine detail of the algorithm is not important here, the main focus of attention being how Shieber modifies it to handle ID/LP directly. Earley's algorithm maintains tables of partially constructed constituents ('items') similar to the 'active edges' of a chart parser. An example of an item might be (9).

(9)   [A → B C.D E, 3]

This item represents an intermediate state in an attempt to build an instance of an A using the CF rule $A → B C D E$ starting at the beginning of word 3 in the input phrase. The position of the period in the daughter list indicates what stage in the construction of A this item represents: categories preceding the period (B and C in the example) have been successfully located, those following it (C and D) have yet to be found. Such an item is stored in a table of all items ending at a particular point in the input phrase, in this case the right end of the C category that has been found.

Items are created in three ways. Firstly, items predicting the distinguished symbol of the grammar are created explicitly in the initialisation stage of the algorithm. Secondly, new items are created to predict the presence of categories needed by unfinished items already in the tables. The next category needed by the example item is a D, so the algorithm predicts a D by adding an item for each grammar rule in which D is the mother category. In both these sorts of items, the period is put to the left of the leftmost daughter, signifying that no daughters have been located yet.

Finally, items are created by combining two existing items together to form a new one. An item is *complete* if its period is to the right of the rightmost daughter (signifying all the daughters have been found). Such a complete item represents the successful application of the grammar rule contained in the item and hence the location of a new category. When such an item is added to a table, a check is made in the table of items finishing where this item starts, to see if any of those items wants to use the new-found category. This check consists of determining, for each item in the table, whether the new-found category is the same as the one immediately to the right of the period (D in (9)). If so, a new item representing the combination of the two items is created, such as (10), and added to the table of items ending where the new category ends.

(10)   [A → B C D.E. 3]

Shieber's modification has two basic components. First of all, the CF rules in the items are replaced by ID rules (where the daughter categories are an unordered set). The item now contains a *sequence* of daughters that have been found and a *set* of daughters yet to be found. Secondly, Earley's algorithm makes use of the 'next sought daughter' in an item (the one immediately to the right of the period) in two places. One is when predicting categories needed by an item already under construction: the category an item needs is its next sought daughter. The other is when combining items together: the new-found category has to be the next sought daughter of the incomplete item for combination to take place. In both these cases, Shieber's algorithm is faced with a *set* of sought daughters, rather than an ordered sequence, and uses the LP rules to determine a next sought daughter from among them. A next sought daughter is one which may legally (according to the LP rules) precede all the others in the set [5].

Aside from these changes, Shieber's algorithm follows Earley's. In effect Shieber's direct ID/LP parser builds CF rules incrementally from ID rules, based on the actual data located in the input phrase. Initially it knows only the dominance information but as each candidate next daughter category is found, an item is added which instantiates that particular possible daughter ordering option. This procedure results in a saving over the context-free version in cases where two context-free rules have the same daughter set and the same ordering for a leading subsequence of daughters, such as (11), since Shieber's algorithm represents the two cases as one right up to the point where ordering between the two rules diverges.

(11)   A → B C D E
        A → B C E D

The abbreviatory nature of Shieber's algorithm   This is, of course, only a very brief sketch of the algorithm, but it is sufficient to demonstrate the main point which is this: although Shieber's algorithm does use ID and LP components separately, it always tests the LP rules in the context of some ID rule. In other words when it does an LP test between two categories, it knows which ID rule it wants them to be daughters in. This is also true of the other algorithms cited above. This may not seem a very significant point but in fact it is the crucial giveaway to the fact that these algorithms are only using ID/LP as an abbreviation mechanism. The reasoning runs as follows.

A minor change to Shieber's algorithm allows it to parse arbitrary *proto*-ID/LP grammars. Instead of accessing a global set of LP rules, suppose each item is tagged with the LP rules relevant to the particular ID rule in the item. Whenever the algorithm predicts the existence of a new category, it adds not only one new item for each ID rule that expands that category, but one for each occurrence of that ID rule in a different ID/LP couple (in the proto-grammar), tagging each item with the set of LP rules from the couple it is associated with. Thus initially each item contains an ID rule and a self-contained set of LP rules governing the daughter orderings, but different items may have different sets of LP rules (taken from different couples). When two items combine, the item thus created inherits the ID rule set from its incomplete ancestor.

Shieber's algorithm does not assume any relationship between the LP rules applied to one ID rule and those applied to another: it only requires that the same LP rules are always used in the items contributing to the construction of any single instance of an ID rule. Hence this modification to it does not affect the basic mechanism of the algorithm. But it does let it parse proto-ID/LP grammars as well as ID/LP grammars, because the choice of LP rules can be localised to particular ID rule instances.

In summary, the fact that LP rules are applied in the context of a given ID rule in Shieber's parser, allows it to be extended to parse proto-ID/LP grammars as well. Furthermore when applied to an ID/LP grammar, the modified algorithm behaves exactly as Shieber's does. Since there is only one ID/LP couple in the grammar, all the items are tagged with the same set of LP rules. The only difference between this and Shieber's version is that in the latter, this set of ID rules is a constant, rather than being specified repeatedly in the items.

In other words, Shieber's algorithm is just an algorithm for parsing proto-ID/LP grammars directly, applied to ID/LP grammars as a special case of proto-ID/LP grammars. But no proto-ID/LP grammar parser can exploit the explanatory value of ID/LP, since the proto-ID/LP formalism does not possess it: proto-ID/LP grammars are essentially *purely* an abbreviatory device, a sometimes more compact representation for arbitrary context-free grammars. Thus Shieber's parser cannot be exploiting this explanatory value either.

Exploiting the ID/LP formalism   Having established that the existing algorithms for ID/LP parsing do not exploit the explanatory potential of the formalism, clearly the immediate task is to propose an algorithm that does. The feature of ID/LP not captured by Shieber's algorithm is the fact that the LP rules are entirely independent of the ID rules; the ordering they impose on daughter categories is globally defined. This means that one can ask whether a sequence of categories is well-ordered without reference to any ID rule. It is this property that distinguishes ID/LP grammars from proto-ID/LP grammars, and it is this property that I shall exploit. The algorithm presented below manipulates the LP rules entirely independently of the ID rules, attempting to build up arbitrary legally ordered sequences of categories before considering ID rule application [6].

The algorithm is based around a conventional chart parser, as described in (Kay 1980, Thompson 1981 etc.). However, the arcs in the chart correspond not to complete and partial constituents, but to legally ordered sequences of (complete) constituents. Each arc is labelled with a multiset (in the sense of (Klein 1983, Gazdar *et al* 1985 p.53ff)) of categories, and signifies that the substring that it spans can be viewed as a legal ordering of those categories. The chart is initialised with an arc for each lexical category and parsing proceeds by looking for interactions between adjacent arcs in the chart, some of which generate new arcs to be added. The conditions under which a pair of adjacent arcs interact are given in (12).

(12)   a)   the right-hand arc is labelled with a single category, and
       b)   the LP rules do not require the category labelling the right-hand arc to precede any arc in the label of the left-hand category.

When two arcs interact, they generate a new arc labelled with the multiset union of their labels. This arc is scheduled for addition to the chart. In addition, whenever a new arc is added to the chart, its label is compared with the ID rule daughter multisets: if ever it matches one, a new arc labelled only with the mother of the ID rule is scheduled. As usual, processing continues until no new arcs can be added In the final chart, any arc spanning the whole input string which is labelled with a single category [7] corresponds to a successful parse as an instance of that category.

For example, the following is a brief commentary of the algorithm in operation, using the simple ID/LP grammar given in (13) (a) on the sentence 'Mary sent John upstairs'. The final chart is given in (13) (b).

(13)   a)   s → np, vp
            vp → v, pp, np

            v < {np, pp} < vp

       b)   Mary sent John upstairs
            A   -np- --v- -np- ---pp---
            B   ---v,np--
            C   ----np,pp----
            D   ------v,np,pp-----
            E   --------vp--------
            F   ----------np,vp--------
            G   -----------s----------

A introduces the lexical arcs. The initial *np* will not combine with the *v*, since the LP rules require *v* < *np*, but the *v* and the second *np* combine, giving B, and the *np* and *pp* (whose relative order is not constrained) combine giving C. B combines with the *pp*, giving D, since the *pp* need not precede either the *np* or the *v*. C does not combine with the *v*, however, because the right-hand arc must have only one category in its label (this constraint blocks multiple derivations of complex arcs). D matches the daughter set of the *vp* ID rule, so E gets added, and that in turn combines with the initial *np* giving F. This matches the *s* rule daughters, giving the final arc, G, which represents a parse of the string as an *s*.

DISCUSSION

The algorithm proposed above is expository rather than efficient. Nevertheless there are some comments one can sensibly make about its performance. In Shieber's parser, and even more so in a context-free parser, a freely ordered language tends to lead to a proliferation of arcs, because so many combinations of categories are legal and the parser has to keep track of each and the ID rules associated with them. The parser introduced here avoids the overheads of incrementally building constituents by using the LP rules to filter candidate sequences of categories before presenting them to the

ID rules. Thus the performance of the parser partly depends on how well the LP rules correctly pick out ID rule daughter sets. The parser will perform well on grammars in which the category ordering information alone provides a useful handle on constituent structure; in other words grammars in which the fact that two adjacent categories in a phrase are allowed to be adjacent (by the LP rules) means it is likely they are part of the same constituent. Put the other way round, the cases where LP rules fail between adjacent categories should ideally be good indicators of higher level constituent boundaries. If this were so for natural language grammars, then one might look for a levelling off of performance between languages like English which have quite strict ordering constraints, and languages like Makua (see (Gazdar and Pullum 1981, Stucky 1981)) which do not, when compared with conventional algorithms: in the new parser the complexity introduced by the flexibility in ordering is not compounded with ID rule book-keeping.

However, as discussed in more detail in (Evans *forthcoming*), this turns out not to be the case: free-order grammars still fare significantly worse. But attention is drawn to the reason for this failure. The LP rules are doing the right job (basically, detecting candidate constituent boundaries to make ID daughter set tests a reasonable proposition) but they are not doing it well enough. Too often, constituent boundaries occur between categories about which the LP rules say nothing at all. To this parser, 'no LP rule' means 'no ordering constraint', and so the categories are legally ordered and can be combined. But in actual fact, 'no LP rule' can also mean (and perhaps more often means) 'these two categories *never* co-occur as sisters, so there is no need to say anything about their ordering'. In such cases, the parser would be far better off *not* combining the arcs.

(Evans *forthcoming*) presents a modification to the algorithm that takes this factor into account. A new component is introduced into the formalism: a set of *sisterhood rules* which make statements of the form 'A and B are allowed to be sister categories'. These are stated separately from the LP rules, but contribute in a similar fashion to the condition on extending an arc in the chart. Equipped with this additional information, far better performance is obtained from the new parsing algorithm, and a levelling off of performance is more evident. Thus the motivation for the algorithm is valid, but the syntactic information required to put it to practical use is not fully represented in a conventional ID/LP grammar.

This new sisterhood component of the grammar also has theoretical utility, enabling a range of additional syntactic generalisations to be captured. This particular aspect of the proposals here is significant: it is only because the computational position adopted is on an equal standing with the theoretical one (rather than somewhere between the high level theoretical representation and the intermediate context-free representation, as existing algorithms are) that one can reasonably expect *any* constructive feedback from the computational domain to the theoretical domain.

The reader is directed to (Evans *forthcoming*) for the full detail of these developments, as well as several additional examples of high level direct interpretations of components of GPSG. The lesson here is that although the algorithm presented above is only a toy one, it not only demonstrates the basic principle of exploitation of the explanatory side of GPSG, it also shows some of the advantage in so doing. This approach advocates development of a wide range of more complex algorithms based on the high level syntactic properties of GPSG, rather than just the uniform simplicity of context-free grammars. In so doing, the door is opened for a richer interaction between computational and formal insight into the nature of syntactic structure.

FOOTNOTES

1. See (Gazdar *et al* 1985) etc. Contrary to appearances, detailed knowledge of the GPSG formalism is not a prerequisite for reading this paper.

2. ID rules are the basic phrase definition rules in a GPSG grammar. I shall discuss them in more detail below, but meanwhile the distinction between 'ID rule' and 'context-free rule' is negligible in this example.

3. (4) (a) alone, however, does not: the order the categories in the daughter set are written down in is arbitrary, the expression A → C,D,B denotes the same ID rule.

4. Only those grammars satisfying the ECPO property of (Gazdar and Pullum 1981) can be put into ID/LP format.

5. Since LP rules define only a partial ordering on categories, there may be more than one next sought daughter in a given set, introducing a further element of non-determinism into the algorithm.

6. However, please note that, as I shall discuss further below, this algorithm is expository rather than efficient.

7. The distinguished category if the grammar has one.

REFERENCES

Barker, C., 1986, "A GPSG Parser in LISP," Unpublished Paper, The Syntax Research Center, University of California, Santa Cruz, Ca..
Earley, J., 1970, "An efficient context-free parsing algorithm," in *Readings in Natural Language Processing*, ed. Barbara J. Grosz, Karen Sparck Jones and Bonnie Lynn Webber, pp. 25-33, Morgan Kaufmann, Los Altos, 1986.
Evans, R., 1982, "A Model of English Verbs," Unpublished Paper, University of Sussex.
Evans, R., 1985, "ProGram - a development tool for GPSG grammars," *Linguistics*, vol. 23, no. 2, pp. 213-243.
Evans, R., (forthcoming) "Theoretical and Computational Interpretations of Generalised phrase Structure Grammar," D.Phil Thesis, University of Sussex.
Evans, R., and Gazdar, G., 1984, *The ProGram Manual*, Cognitive Studies Research Paper 035, Cognitive Studies Programme, University of Sussex, Brighton.
Gazdar, G., Klein, E,. Pullum, G. and Sag, I., 1985, *"Generalized Phrase Structure Grammar*, Blackwell, Oxford.
Gazdar, G. and Pullum, G., 1981, "Subcategorization, constituent order and the notion "head"," in *The Scope of Lexical Rules*, ed. M. Moortgat, H. v. d. Hulst and T. Hoekstra, pp. 107-123, Foris Publications, Dordrecht.
Kay, M., 1980, "Algorithm Schemata and Data Structures in Syntactic Processing," CSL-80-12, Xerox PARC, Palo Alto.
Kilbury, J., 1984, "Earley-basierte Algorithmen fur Direktes Parsen mit ID/LP-Grammatiken," KIT-Report 16, Technische Universitat Berlin, Berlin.
Klein, E., 1983, "A Multiset Analysis of Immediate Dominance Rules," Unpublished Paper.
Pulman, S., 1983, "Generalised phrase structure grammar, Earley's algorithm, and the

minimisation of recursion," in *Automatic Natural Language Parsing*, ed. Karen Sparck Jones and Yorick Wilks, pp. 117-131, Ellis Horwood/Wiley, Chichester/New York.

Shieber, S., 1984, "Direct parsing of ID/LP grammars," *Linguistics and Philosophy*, vol. 7, pp. 135-154.

Stucky, S., 1981, "Word order variation in Makua : a phrase structure grammar analysis," PhD Dissertation, University of Illinois, Urbana-Champaign.

Thompson, H., 1981, "Chart parsing and rule schemata in PSG," *ACL Proceedings, 19th Annual Meeting*, pp. 167-172.

Search Control
Techniques

STATISTICAL HEURISTIC SEARCH

Bo Zhang and Ling Zhang*

Dept. of Computer Science, Tsinghua University, Beijing, China
*Dept. of Mathematics, Anqing Teachers'College, Anqing, Anhui, China and The Institute of Intelligent Machines, ACADEMIA SINICA, Hefei, Anhui, China

ABSTRACT

Under certain hypotheses a heuristic search can be considered as a random sampling process. Thus, it is possible to transfer the statistical inference method to the heuristic search.
Based on the idea above, in [1] - [4] we have discussed some new algorithms which incorporated a specific statistical inference method into some heuristic search procedures.
In this paper we summarize the general principle of the new algorithms and present the main results of the combining those two techniques.

Index: heuristic search, statistical inference, computational complexity

INTRODUCTION

A. Heuristic Search

Heuristic search is a search procedure which uses heuristic infomation. It is one of the basic techniques in computer problem solving and has been investigated by many researchers [5] - [12]. For example, Nilsson and Hart et al. [14] presented algorithm A* and its several properties. Then Pearl et al. [5] - [8] made a thorough study from probabilistic point of view about the relations between the precision of the heuristic estimates and the average complexity of A*.

Pearl assumes the following probabilistic search space: a uniform m-ary tree G has a unique goal node S_N at depth N at an unknown location. The algorithm A* searches for the goal node S_N using the evaluation function
$$f(n)=g(n)+h(n)$$
where $g(n)$ is the depth of node n and $h(n)$ is a heuristic estimate of $h*(n)$, the distance from n to S_N. The estimates $h(n)$ are assumed to be random variables ranging over $[o, h*(n)]$, characterized by distribution function $F_{h(n)}(x)=P[h(n) \leqslant x]$. $E(Z)$, the expected number of nodes expanded by A*, is called the mean complexity of A*.

One of his results is that if the heuristic estimate $h(n)$ satisfies

$$P\left[\frac{h*(n)-h(n)}{h*(n)} > \varepsilon\right] > \frac{1}{m}, \quad \varepsilon > 0$$

where m is the branching factor, then $E(Z) \sim 0(e^{CN})$, $C > 0$.

*This work was supported in part by the AcademicFoundation of China.

Thus, in general, the "exponential explosion" cannot be complete-ly overcome in A*. Its search efficiency is rather low, so are the other heuristic search algorithms.

B. Statistical Inference

Statistical inference is an inference techinque for testing some statistical hypothesis-an assertion about a distribution of one or more random variables based on their observed samples. It is one major area in mathematical statistics [13][16].

For example, in the Wald sequential probability ratio test(SPRT) for two hypotheses, assume that $x_1, x_2, \ldots, x_n, \ldots$ are independent and indentically distributed (i. i. d.) random variables. $f(x; \mu)$ is its density distribution function. Two simple hypotheses are $H_0: \mu = \mu_0$, $H_1: \mu = \mu_1$, $\mu_1 \neq \mu_0$. Thus, given n observations, we form the sum

$$S_n = \sum_{i=1}^{n} \ln \frac{f(x_i; \mu_1)}{f(x_i; \mu_0)} \qquad n \geqslant 1.$$

The stopping rule says that sampling is continued as long as $-b < s_n < a$.

If sampling terminates with R observations, hypothesis H_0 is accepted if $S_R \leqslant -b$, and H_1 is accepted if $S_R \geqslant a$. Where $o < a < b < \infty$, a and b both are given constants.

Conclusion 1: If hypothese H_0 and H_1 are true, the stopping variable R is finite with probability one.

Conclusion 2: If $P_\mu(|Z| > o) > o$, then $P_\mu(R > n) \leqslant e^{-Cn}$, $C > 0$. Where R is the stopping variable in SPRT.

Conclusion 3: Given a significance level (α, β), let

$$A = \frac{1-\beta}{\alpha}, \quad B = \frac{\beta}{1-\alpha}, \quad Z_i \triangleq \ln \frac{f(x_i; \mu_1)}{f(x_i; \mu_0)}, \quad Z \triangleq \ln \frac{f(x; \mu_1)}{f(x; \mu_0)}.$$

$E_{u_i}^{\wedge} |Z| < \infty$, $E_{\mu_i}^{\wedge} Z \neq 0$ $(i = 0,1)$, the mean of stopping variables (sample size) in SPRT is

$$E_{\mu_i}^{\wedge}(R) \approx \frac{\alpha \ln \frac{1-\beta}{\alpha} + (1-\alpha) \ln \frac{\beta}{1-\alpha}}{E_{\mu_i}^{\wedge}(Z)} \qquad (1)$$

If

$$f(x; \mu) = \frac{1}{\sigma \sqrt{2\pi}} \exp\left[-\frac{1}{2}\left(\frac{x-\mu}{\sigma}\right)^2\right]$$

then

$$Z \triangleq \ln \frac{f(x; \mu_1)}{f(x; \mu_0)} = \frac{1}{2\sigma^2}\left[2x(\mu_1 - \mu_0) + (\mu_0^2 - \mu_1^2)\right]$$

$$S_n \triangleq \sum_{i=1}^{n} Z_i = \frac{1}{2\sigma^2}\left[2(\mu_1 - \mu_0) \sum_{1}^{n} x_i + n(\mu_0^2 - \mu_1^2)\right]$$

The stopping rules of SPRT are as follows.

If $\sum_{i=1}^{n} x_i \geqslant \frac{\sigma^2 g_1}{\mu_1 - \mu_0} + \frac{n}{2}(\mu_1 + \mu_0)$, hypothesis H_0 is rejected.

If $\sum_{i=1}^{n} x_i \leqslant \frac{\sigma^2 g_2}{\mu_1 - \mu_0} + \frac{n}{2}(\mu_1 + \mu_0)$, hypothesis H_0 is accepted.

Otherwise, observing x_{n+1} is continued.

where
$$g_1 = \ln \frac{1-\beta}{\alpha}, \quad g_2 = \ln \frac{\beta}{1-\alpha} \tag{2}$$

The error probability of type I, $P_1 \leqslant \alpha$ (H_0 is rejected when H_0 is true).
The error probability of type II, $P_2 \leqslant \beta$ (H_1 is accepted when H_0 is true).

C. Statistical Heuristic Search

In heuristic search, given a graph G and an evaluation function f(n), when a new node has been expanded, a set of values f(n) of its successors is computed. According to a given expanding rule of nodes (in A or A*, a node having a lower evaluation function will be expanded first), the node being expanded next is decided ......, The expanding process is continued until a goal is found.

In statistical inference (SPRT), given a random variable X and a statistic a(n), when a new sample has been chosen, its a(n) is computed and statistical inference is exercised. According to a given stopping rule, observation is continued until a hypothesis H_0 is accepted or rejected.

Based on the similarity between these two techniques, when we consider a heuristic search as a random sampling process, it is possible to transfer the matured statistical inference technique to the huristic search. Thus, we provide heuristic search with a new tool. Here we use SA instead of SA* in [1][3].

ALGORITHM SA

A. The Model of Search Tree

Assume search tree G is a uniform m-ary tree having an initial node S_0 (root) and a unique goal node S_N at depth N such that
$$v(S_N) \leqslant v(p)$$
p ∈ the nodes at depth N
(If $v(S_N) < v(p)$

Fig.1

p ∈ the nodes at depth N and $p \neq S_N$, then S_N is a unique goal).
For any node n ∈ G, T(n) is a subtree rooted at n (Fig.1). If n is a node at i-th depth, T(n) is called an i-subtree.

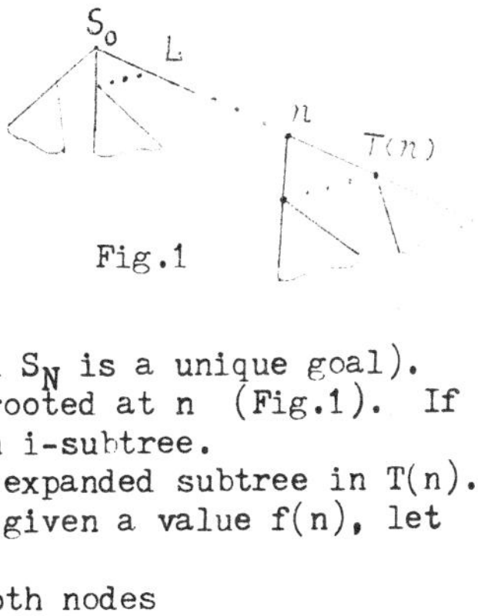

Assume in some search stage, G' is an expanded subtree in T(n).

Heuristic Information: For any n ∈ G, given a value f(n), let f(n) be an estimate of { min (v(p))}.
                    p ∈ T(n), p ∈ N-th depth nodes
Thus in algorithm A (or BF), the open node having the lowest value of f(.) will be expanded first.

B. Statistic a(n)

For applying the statistical inference method, it is necessary to extract an appropriate statistic from f(.). There are many means available. In the following discussion we introduce one of the possible ways.

Fixed n ∈ G, let $T_k(n)$ be the expanded tree in T(n) having k nodes. Let
$$a_k(n) = F(f(p), \ p \in T_k(n)) \tag{3}$$

Where F is a combination function of $f(p$
When a node is expanded in $T(n)$, from (3) a new $a_k(n)$ is obtained. We say "observing $T(n)$ is continued". It means expanding node
in $T(n)$ and computing a new statistic $a_k(n)$, called a new observation, from (3). We say "exercising some statistical inference over T".
It means exercising some statistical inference over the statistics
$a_k(n)$ corresponding to $T(n)$.
For $\{a_k(n)\}$ we now make the following assumption.
Hypothesis I: For any $n \in G$, assume that $\{a_k(n)\}$ is an i. i. d.
random variable. Let L be the shortest path from $S_0 \to S_N$, when $n \in L$,
have $\mu(n)=\mu_0$; $n \notin L$, $\mu(n)=\mu_1 > \mu_0$, where $\mu(n)$ is the mean of $\{a_k(n)\}$.

C. Algorithm SA
   See [2] for details.

D. Algorithm SPA and Its Complexity
   In the preceding section, we presented the general principle of
SA. As an example, we now use SPRT as a specific testing hypothesis
for unraveling the characteristics of SA.
   Definition: If in SA as testing hypothesis S, SPRT is exercised

over m i-subtrees, using a level ($\alpha_i = \dfrac{\alpha}{(i+1)^2}$, $\beta_i = \dfrac{\alpha}{(m-1)(i+1)^2}$),
and when the sample size surpasses a given threshold d_i (the value
d_i will be given below), the hypothesis H_0 is rejected, $i=0,1,2,\ldots$,
we define this SA as SPA under level ($\alpha, \dfrac{\alpha}{m-1}$), denoted by SPA for
short.
   Let $\{a_k(n)\}$ be $\{x_k\}$, having an $N(u,\sigma^2)$ distribution. Given two
simple hypotheses $H_0 : \mu = \mu_0$, $H_1 : \mu = \mu_1$, $\mu_1 > \mu_0$. By substitut-
ing $\dfrac{\alpha}{(i+1)^2}$ for α and $\dfrac{\alpha}{(m-1)(i+1)^2}$ for β in (1) and taking $d_i = 2b_2$

$\ln(i+1).\ln(\dfrac{i+1}{\alpha})$, $b_2 = \dfrac{4m\,\sigma^2}{(\mu_1 - \mu_0)^2}$, we have.

   Theorem 1: Given α_0, β_0. Let $\alpha = \min(\dfrac{\alpha_0}{A}, \dfrac{\beta_0}{A})$, $A = 2\sum_{i=1}^{\infty}\dfrac{1}{i^2}$.
   If $\{a_k(.)\}$ has a normal distribution, using SPA, under level (α ,
$\dfrac{\alpha}{m-1}$), the upper bound of the complexity of finding a solution path in
G is $O(N. Ln^2 N)$. The error probability of type, I, $P_1 \leqslant \alpha_0$. The error
probability of type II, $P_2 \leqslant \beta_0$.
   Proof: In i-th depth, the threshold

$$d_i = 2b_2\ Ln(i+1).\ln(\dfrac{i+1}{\alpha}) \sim C\ ln^2(i+1)$$
So the upper bound of the complexity is

$$\sum_{i=0}^{N} C\ ln^2(i+1) \sim O(N.Ln^2 N).$$

If in the search process the sample size has never surpassed the
threshold, from $(1)(2)$, it is noted that deciding on m i-subtrees,

the error probability of type I $\leqslant \dfrac{\alpha}{(i+1)^2}$, So the total error

$$P_1 \leqslant \sum_{i=0}^{N} \frac{\alpha}{(i+1)^2} \leqslant \alpha \sum_{i=1}^{\infty} (\frac{1}{i})^2 = \frac{\alpha A}{2} < \frac{\alpha_0}{2}$$

In some search stage, if the sample size surpasses the threshold and H_0 is rejected, the error probability des not change if the subtree being searched does not contain the goal, and the error probability will increase if the subtree contains the goal. We estimate the increment as follows.

From Conclusion 2, the distribution of the stopping variable R in SPRT

$$P(R > n) \leqslant e^{-Cn} = \int_{n}^{\infty} Ce^{-Cx} \, dx$$

Assume $P(R > n) = e^{-Cn}$ (4)

From Level $(\frac{\alpha}{(i+1)^2}, \frac{\alpha}{(m-1)(i+1)^2})$ at i-th depth, the mean of R is

$b_2 \ln(i+1)$, $b_2 = \frac{4m\sigma^2}{(\mu_1 - \mu_0)^2}$. And from (4)

$$E(R) = \int_{0}^{\infty} Cx \cdot e^{-Cx} \, dx = \frac{1}{C}$$

Thus, $C_i = \frac{1}{b_2 \ln(i+1)}$, the value of C at i-th depth.

The probability that the sample size surpasses the threshold at

i-th depth is $P(R > d_i) = e^{-C_i d_i} = \frac{\alpha^2}{(i+1)^2} < \frac{\alpha}{(i+1)^2}$. i.e., When the

sample size surpasses the threshold, the rejection of H_0 will

cause $< \frac{\alpha}{(i+1)^2}$ increment of the error probability of type I.

Finally, $P_1 \leqslant \sum_{i=0}^{N} \frac{2\alpha}{(i+1)^2} \leqslant \alpha A = \alpha_0$.

Similarly, $P_2 \leqslant \beta_0$.

Certainly, when the sample size surpasses the threshold, the rejection of H_0 does not change the error probability of type II.

Corollary 1: Assume that $\{a_k(.)\}$ has a distribution function

$f(x ; \mu)$. Let $Z \triangleq \frac{f(x ; \mu_1)}{f(x ; \mu_0)}$, $E_{\mu_i}^{\wedge} |Z| < \infty$, $E_{\mu_i}^{\wedge} Z \neq 0$, $i = 0, 1$. The

upper bound of the complexity of SPA is $O(N \cdot \ln^2 N)$.

Successive SA

We summarize the above results as the following. Under a given level (α, β), the SA search results in the polynomial complexity C $(N \cdot Ln^2 N)$. Unfortunately, a real goal can only be found with probability $(1-b)$, $b = \alpha + \beta$. The deficiency can be overcome by using SA search successively.

We imagine that if a node at depth N found by the SA search is not a goal of G, algorithm SA is applied to the remaining part of G once again. Thus, the probability of finding a real goal is increased by $b(1-b)$ or error probability is decreased to b^2, ..., the repeated use of SA is continued until the goal is found. We call this procedure successive SA or SA for short.

Using the same proof as in [3] , we have

Theorem 2: In a uniform m-ary tree, using the SA search, a real goal can be found with probability one, and the upper bound of the complexity remains C_3 (N. $\ln^2 N$).

GRAPH SEARCH

A. General (OR) Graph Search

Algoritm SA can formally be extended to graph-search.

First, graph-search needs to be transferred into some sort of tree search. There are several strategies dealing with the problem. For example, the procedure presented in [14] is one of the stratigies. It generates an explicit graph G and a subset, T, of graph G called the search tree (see [14] for more details).

Second, since depth N where a goal is located is ussually unknown, the branching factor is not a constant, and there is not only one solution path etc.. There is no threshold to be given in the graph-search algorithm.

Finally, because N is unknown, if all solution paths were to be pruned off (although its probability is quite low), the SA search may not terminate. To solve this problem, we must set up an upper bound B-the estimate of depth N. When reaching bound B, stop the currently executed SA search, and start a new round of search.

As mentioned before, the depth that the goal is locate is usually unknown, in reality, the statistic constructed does not satisfy i. i. d. perfectly etc.. So it is difficult to estimate the complexity of graph-search theoretically.

We have taken 8-puzzle as an experimental model to compare WSA (weighted SA search) [4] with A. In 80 instances, WSA is superior to A with 90% and the mean decrement in the complexity is about 25-30%.

In these experiments, the problem scale is not very large. The longest solution path does not exceed 35 steps. Even though for such medial scale problems as 8-puzzle, SA is dominant to A with a high probability. We may gain much profit from using SA in solving large scale probelem.

B. AND/OR Graph Search

Generally, statistical inference methods may be extended to AND / OR graph search.

In AND/ OR graph search the objects of pursuit are solution graph. Each such subset is called a solution base and the one-to-one correspondence between a solution base and a given node no longer holds. So the translation of the statistical inference method to the AND/OR graph search is slightly different.

As an example, we assume the statistical inference method is SPRT and the AND/OR graph search algorithm is GBF [15] .

Assume G(n) is a subgraph of G rooted at n and $G_k(n)$ is the expanded portion of G(n) having k nodes. For any $q \in G_k(n)$, let f_1 (q) be the graph evaluation function of the solution base contain-ing node q. So $f_1(q)$ estimates some properties of the set of solution graphs that may emanate from the given candidate base B(n).

Define $a_k(B(n))=F(f_1(q), q \in G_k(n))$.

Where F is a combination function of $f_1(q)$.

Assume there exists a unique optimal solution graph G_0 in G. For any solution base B(n) containing n, while expanding $\hat{G}(n)$ rooted at n for searching G_0, the observation $\{a_k(B(n))\}$ taken from expanded subgraph $G_k(n)$ of $\hat{G}(n)$ estimates the promise offered by the expansion of $\hat{G}(n)$. So $\{a_k(B(n))\}$ can be used as a statistic of the SPRT to identify the most promising solution base. Thus, a new statistical AND/OR graph search algorithm SGBF is obtained.

Assume G is a general (2N, b)-game tree having a unique optimal solution graph (goal). The statistic $\{a_k(n)\}$ satisfies an asumption similar to Hypothesis I. Use the SPRT as a testing hypothesis S. At the 2n-th depth, let the significance level be

$(\alpha_i = \dfrac{\alpha}{b^n(n+1)^2}, \beta_i = \dfrac{\alpha_i}{b-1})$, α is a given constant, $0 < \alpha < 1$, and the

threshold $E = Cn^2$, where C is a constant independent of n. The complexity of an algorithm is defined as the expected number of frontier nodes examined by the algorithm [15]. We have.

Theorem 3: When SGBF searching the (2N , b)-game tree G, the goal can be found with probability one and the upper bound of the complexity is O $(N^2 \cdot b^N)$.

Theorem 4: When SGBF searching a two-player (2N,b)-game tree, if there exists a winning strategy for MAX, at the 2n-th depth let the threshold $E = C \cdot \ln^2 n$, C is a constant, then MAX can force a win with probability $(1-\alpha)$, $0 < \alpha < 1$. And the upper bound of the complexity is O $(N \cdot \ln^2 N)$.

The proof of the theorems above is similar to Theorem 1 and 2.

OTHER STATISTICAL INFERENCE METHODS

We only deal with the SPRT in the preceding discussion. Generally, the complexity of SPRT is lower than others, but the parameters α^2, μ_1 and μ_0 are usually unknown. We may use

$S_n = \dfrac{1}{n-1} \sum\limits_{i=1}^{n} (x_i - \bar{x})^2$, $\bar{x} = \dfrac{1}{n} \sum\limits_{1}^{n} x_i$ to estimate α^2; the minimum of

mean statistics among all i-subtrees to estimate μ_0 ; the average of mean statistics (except the minimum) of i-subtrees to estimate μ_1. But these will raise new errors.

In order to overcome the above difficulty, other statistical inference methods, such as u-test, t-test, may be adopted. And it can be proved that under these methods Theorem 1 also holds.

For example, using u-test to determine the validity of $\mu_1 = \mu_0$, that is, whether the mean of random variable X is equal to that of Y, we take the following composite statistic.

$$U = (\bar{X} - \bar{Y}) \sqrt{\dfrac{\sigma_1^2}{1} - \dfrac{\sigma_2^2}{n}} \qquad (5)$$

Where σ_1 and σ_2 are known, $\bar{X}$ and $\bar{Y}$ are the observed sample mean of X and Y, 1 and n are the sample size of X and Y, respectively.

In tree search, while node p is expanded, it has m successors: p_1 , p_2 , $\ldots$, p_m. Let T(p) be a subtree rooted at p. The mean statistics $\bar{X}_i$ corresponding to $T(p_1)$, $T(p_2)$, $\ldots$, $T(p_m)$ are computed.

Assume $\bar{X}_1 \leqslant \bar{X}_2 \leqslant \ldots \leqslant \bar{X}_m$.

We now use u-test to judge whether the mean statistic of $T(p_1)$ is equal to that of $T(p_2)$. Given level (α, β) and sample size (1, n). (1+n) gradually increases in the testing precess, 1+n = 1, 2, ... From $\Phi(k_\alpha) = 1 - \alpha$, we obtain a constant k_α, where $\Phi(.)$ is a standard normal distribution function [16].
If $|U| > k_\alpha$ subtree $T(p_2)$ is pruned off.
$|U| \leqslant k_\alpha$ sample size (1+n) is increased by one, that is, a new node is expanded in $T(p_1)$ or $T(p_2)$ by algorithm A.
The process continues until a goal is found. The composite statistic U is obtained from (5) by subtituting $\bar{X}_1$ for $\bar{X}$ and $\bar{X}_2$ for $\bar{Y}$. We also take threshold $D(k) \sim O(\ln^2 k)$ in k-th depth, when sample size $\geqslant D(k)$, subtree $T(p_1)$ is accepted.
It can be proved that its complexity is of the same order of magnitude as algorithm SPA.
If σ_1 and σ_2 are also unknown, t-test may be used.

CONCLUSIONS

Under certain conditions, statistical inference technique can be introduced to heuristic search so that various kinds of statistical heuristic search algorithm can be constructed.
For some statistical inference methods and heuristic searches, we prove that the combination of these two techniques will result in an improvement on searching efficiency.

REFERENCES

[1] Zhang, Ling and Zhang, Bo, 1983, The Statistical Inference Method in Heuristic Search Technique, Proc. of 8-th IJCAI-83, 757-759.
[2] Zhang, Bo and Zhang, Ling, 1985 a, A New Heuristic Search Technique-Algorithm SA, IEEE Trans., Vol. PAMI-7, 1, 103-107.
[3] Zhang, Ling and Zhang Bo, 1984, The Successive SA* Search and Its Computational Complexity, Proc. of 6-th ECAI-84, 249-258.
[4] Zhang, Bo and Zhang, Ling, 1985 b, A New Weighted Technique in Heuristic Search, Proc. of 9-th IJCAI-85, 1037-1039.
[5] Pearl, J. 1981, Heuristic Search Theory: Survey of Recent Results, in Proc. 7th IJCAI-81, 554-562.
[6] Pearl, J. 1980 a, Probabilistic Analysis of the Complexity of A*, Artificial Intell., 15, 3, 241-254.
[7] Pearl, J. 1980 b, Asymptotic Properties of Minimax Tree and Game -Searching Procedure, A. I., 14, 2, 113-138.
[8] Pearl, J., 1983, Knowledge Versus Search: A Quantative Analysis Using A*, A. I., 20.
[9] Baudet, G. M. 1978, On the Branching Factor of the Alpha-Beta Pruning Algorithm, A. I., 10, 173-199.
[10] Knuth, D. E. and R. W. Moore, 1975, An Analysis of Alpha-Beta Pruning, A. I., 6, 293-326.

[11] Kanal, L. and Kumar, V. 1981, Branch & Bound Formunation for Sequential and Parallel Game Tree Searching: Preliminary Results, in Proc. 7th IJCAI, 569-571.

[12] Berliner, H. 1982, The B* Tree Search Algorithm: A Best-First Proof Procedure, in Readings in A. I., Webber, B. L. Ed., 79-87.

[13] Zacks, S. 1971, The Theory of Statistic Inference, New York: Wiley.

[14] Nilsson, N. J., 1980, Principles of Artificial Intelligence, Palo Alto, CA: Tioga.

[15] Pearl, J., 1984, Heuristic, Intelligent Search Strategies for Computer Problem Solving, Addison-Wesley Publishing Company.

[16] Hogg R. V. et al, 1977, Probability and Statistical Inference, Macmillan Publishing Co., Inc.

ON TRYING TO DO DEPENDENCY-DIRECTED BACKTRACKING
BY SEARCHING TRANSFORMED STATE SPACES (AND FAILING)

Sam Steel
Computer Science Dept, University of Essex,
Colchester CO4 3SQ, England

ABSTRACT: Any search involves choices. Bad choices can cause disaster. Dependency-directed backtracking (DDBT) is an attempt to undo only those choices that caused the disaster. One approach to it is to transform the search space of the original problem into an equivalent space with different states and operators which is easier to search.

I show the merits and failings of various spaces. At the moment I have no perfect method.

Keywords: dependency-directed backtracking, search

1. Introduction

There are lots of problems in AI which can be formulated as state space search. The problems that make dependency-directed backtracking (DDBT) worth having can arise in any of them.

What I want to show is that solving a problem can be done by searching any of several different spaces, some simple, some complex; and that DDBT is a hard or impossible strategy in some simple spaces, but an easy strategy (ultimately, almost depth-first with chronological backtracking) in a more complex space. But there is a problem; such strategies find the same solution more than once. I shall do this by going through a sequence of state spaces, from simple to complex, pointing out where some related work on DDBT fits in.

There are at least four traditions now running in DDBT:

--- Work following Phil Hayes (Hayes 1973, 1975) such as (Daniel 1977);

--- Reason maintenance - work following (most notably) Doyle (1979), which is quite close in spirit to the first;

--- Work strongly directed to assisting backtracking in PROLOG-like languages, and therefore much concerned with specialized techniques: eg (Bruynooghe & Pereira 1984), (Dilger & Janson 1986);

--- Assumption based reason maintenance, stemming from eg (de Kleer 1986);

This paper is in the first tradition, with glances at the second. Of course it would be ideal to show how they are all related, but I am unable to do that. The attraction of the first and second is that they are concerned with the study of and-or graphs, an extremely well-known and widely used idea.

State spaces are presented as trees, perhaps with infinite branches, rather than as graphs, perhaps with loops.

Examples are from the domain of planning.

2.   Space 1: the space of domain states

In this space, states represent states of the domain in which the solution is being sought. If the domain is chess, the states are board states. This is standard.

Planning can of course be seen as search of a space whose states are with partially specified plans. The arcs of such a space are plan-improving actions - adding new plan-steps to achieve other plan-steps' preconditions, adding ordering information to prevent destructive interference between plan-steps, and so on.

States will have structure. Some substructures will count as defects that must either be made harmless or else removed. Call such substructures "flaws". There may be many of them in a state.

Making flaws harmless involves adding more structure - for instance, the flaw of an unsupported precondition in a plan can be cured by adding another action that has that precondition as its effect. Call the structure one can add a "fix". A flaw can have zero or more fixes. Fixes can add flaws as well as removing them - for instance, achieving a precondition by adding an action means the new action's own preconditions have to be achieved. Call the curing of a flaw by a fix a "choice". Then each arc in the state space represents a choice.

If a state contains a flaw with zero fixes, the flaw is a failing flaw and the state is a failing state. When this happens, search has to carry on elsewhere in the state space. Where?

The answer to this is part of what distinguishes search strategies. The answer "carry on at an unconsidered child of your most recent ancestor which has unconsidered children" is of course depth-first search with chronological backtracking. The merit of this answer is that to follow it one need not record all of the search space so far considered, just a stack describing the ancestors of the current state and what one has ever chosen when at them. This usually makes vast savings in space. But there are problems with chronological backtracking.

--- Pointless work is done. Alternatives are examined in the tree beneath the point where the problem that caused the failure first arose. Until that problem is removed, alternative fixes to other flaws are useless.

--- Good work is lost. Several problems that arose before the choice causing the failure may have been fixed after that choice. Going back to that choice chronologically will get rid of the useful work too.

3.   Space 2; the space of domain/dependency-graph states

One way of avoiding those problems is to associate with every state a list of flaws extant in it. Any improvement in the state involves choosing some flaw and some fix for that flaw. The improved state should have associated with it a revised list of flaws (the original list, less the the the flaw fixed, plus any new flaws the fix introduced), and associated with the chosen flaw there should also be a record of

--- Which fix was selected to cure the flaw.

--- What changes that fix introduced into the state.

--- What new flaws the fix introduced. The fix is said to be the "parent fix" of these flaws, and the choice of that fix is their "parent choice".

Call such a record the dependency graph. Each state in this new space then has two parts

--- A domain state

--- A dependency graph state

Figure 1 is an example. (The figures are mostly at the end.) w1 and w2 were original flaws, from the statement of the problem. x1.1 and x2.4 are the fixes that were chosen for them in this state. c1.1 and c2.4 are records of whatever change was added to the domain state by those fixes. w3 and w4 are flaws that were introduced by the choice of w1.1 as the fix.

How does this help with backtracking?

--- Pointless work need not be re-done. When a flaw is discovered to fail, the dependency graph displays which fix (if any) introduced that flaw, and which earlier flaw that fix was introduced to cure. Then one should undo that early fix and try again.

--- Good work is kept. When a fix is undone, all flaws of which that fix is an ancestor vanish, and all fixes of such flaws and any changes involved can be forgotten too. But all other fixes, whether made before or after the culprit choice are undisturbed.

The effect of a backtrack is then to go straight away to a state in the space where the failing flaw hasn't arisen, but where good choices made since its arising are also present. For instance, figure 2 is a domain/dependency-graph state space, though only the dependency information is drawn in. State A has the dependency graph of figure 1. w3 is discovered to fail. The choice that introduced it was w1/x1.1. Remove that choice from the dependency graph. Remove from the domain state all the changes recorded in the subgraph being deleted below w1/x1.1. The effect of that is as if search had hopped in the state space along the arrow from state A to state B. As promised, pointless alternatives for w1 have not been considered, and the good choice for w2 has not been lost.

3.1 Hayes

What has been said so far is a re-description of (part of) the DDBT that Hayes proposed in (Hayes 1973, 1975) and which has been rather overlooked in the shadow of reason maintenance. I have reviewed it because Hayes presents his ideas as an algorithm in which plans and the associated dependency information are seen as arranged in linear sequence rather than as being states in a search space.

4. Dependency graphs and the and-or formulation of the problem

4.1 And-or trees

Another way of looking at this dependency information is to see each state of the information as a partial solution of the and-or tree representation of the problem being solved. The flaws are goals, the fixes are methods of reducing those goals to other flaws, or subgoals. The flaws are or-nodes, the fixes are and-nodes. I shall speak of flaws and fixes, not or-nodes and and-nodes. The partial solution subgraph (PSSG) is an alternative representation of the partial domain state, because one can derive the domain state by cumulating the changes mentioned in the dependency graph. So search of the space of partial plans is isomorphic to a search of the space of PSSGs of the and-or tree of the problem. But of course there are other ways of searching and-or trees - for instance, a stack-based depth-first method such as Prolog uses. Why the fuss about state space?

4.2 And-or graphs

The reason is that some problems have an and-or formulation that is not just a tree, but at least a directed acyclic graph. This happens whenever a flaw arises not from a single fix but from several fixes at once, so that it has several parent fixes.

This occurs in planning with so-called "interactions". Figure 3 is an example. The precondition P of Act1 is achieved by the insertion of Act2 which has effect P. The interval between Act1 and Act2 is then a protected range for P. But suppose there is another action Act3, not constrained to occur either before Act2 or after Act1, with effect -P. Then there is a flaw, a possible harmful interaction, between the range and the effect of Act3. This flaw only arises when both the range (the fix of one flaw) and Act3 (the fix of another) are present at once. The dependency graph of this situation is figure 4.

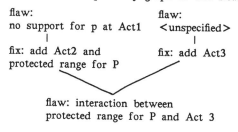

flaw: flaw:
no support for p at Act1   < unspecified >
    | |
fix: add Act2 and   fix: add Act3
protected range for P

flaw: interaction between
protected range for P and Act 3

figure 4

Suppose there is no fix for that interaction flaw. It must be cleared by backtracking. Backtracking to either parent fix would clear it. But unless one is very careful, one loses completeness of search.

4.3 Search in space 2 is not easily both complete and dependency directed

One way of looking at what dependency information provides is to see it as adding pointers from any state in which a flaw X is extant, back to each state just before a choice was made that added one of the parent fixes of X. For example, figure 5 is a space with only the dependency information drawn in. Suppose w4 is a failing flaw, so that state E is a failure state. If w4 had only one parent fix, and so only one back pointer, there would be no problem. As there are two, a choice must be made. Suppose one chooses to get rid of the more recent parent fix. One reaches C. C has no other children, so it too is now a failure state. Backtracking from it takes one to A. But that misses the solution at D. If one had chosen the other parent fix, and gone to B, all would have been well. But there is no record of which parent one backtracked to, so that one can't so to speak backtrack one's backtrack, which is what really needs to be done.

Obviously always going straight to the least recent parent can also miss solutions.

As I understand Hayes' approach, he can easily store dependency information where a flaw has more than one parent, but, once one parent has been chosen to backtrack to, the choice is irrevocable, and so his search is incomplete.

5. Space 3: Solution space

5.1 Solution space

In this space, the states are just states of the dependency graph, which is now interpreted as a PSSG. The domain state is still reconstructible. But the arcs represent a new sort of action, and so the shape of the tree is different. The action is

DO   Take some flaw in the PSSG that currently has no fix, and add a fix to it.

Solution space is redundant, and harder to search, because the order of fixes is unimportant. In figure 6, B and C are the same, thought they are reached differently.

Worse, DDBT is impossible. The only way to get rid of unwanted bits of the current PSSG is to jump about in solution space to states where the unwanted bit has not been added. But jumping about means that all parts of the search space so far found have to be kept. One might as well jump about in the domain space straight away.

5.2 Levi and Sirovich

This space is the space that Levi and Sirovich (1976) proposed as the first method of solving general and-or graphs. Redundancy was to be handled by a search strategy that kept the whole examined search tree, except for those parts below repeated states.

6. Space 4: Solution space with UNDO

Explicit DDBT can be got into something like solution space by adding a new operator, UNDO.

UNDO Take some flaw in the PSSG that currently has a fix, and remove it and any subgraph of which that fix is an ancestor.

Now any choice that is a parent of a failing flaw can be removed exactly. If w4 is a failing flaw in figure 7, it can be removed either by applying UNDO to the fix x1.1, which yields figure ˜8, or by applying UNDO to the fix x2.1, getting a different successor state, as in figure 9. It is the fact that there are these two different states reachable by UNDO that will make DDBT possible when flaws can have several parents. But it also means that this space is even more redundant than pure solution space, both because UNDO can be applied to any flaw whether or not is is a failing flaw, and because any state can be reached by an arbitrary number of DOs and UNDOs alternately adding and removing the same fix for some flaw. The space is in fact a graph, not a tree. But a space that can be searched depth-first must be a tree.

7. Space 5: Decision record space

That redundancy is removed in decision record space, by changing both the space and the search strategy. The states are now not just subgraphs of the and-or graph, but subgraphs with, at each flaw,

--- A current fix: that is, the current choice for solving that flaw. If it is null, the flaw is unsolved.

--- A list of rejected fixes: a list of zero or more fixes that represents choices tried but rejected as ways of solving that flaw. This is the novelty. In drawings, rejected fixes are shown as struck through.

An example of such a state is figure 10. The children of each flaw, taken together, have the right shape to be a decision record about choices made at that flaw. Each flaw has its own decision record. So the space is really a space of sets of decision records. Furthermore, since each set of records is arranged in the right sort of graph, the set also records dependency information.

The actions that link these states are roughly the same as before, except that the DO is sensitive to the list of rejected fixes.

DO   Take some flaw in the PSSG that currently has no fix, and add a fix to it, as long as that fix isn't a rejected fix for that flaw.

UNDO Take some flaw in the PSSG that currently has a fix, and remove it and any subgraph of which that fix is an ancestor.

Because neither DO nor UNDO actions can decrease the lists of rejected fixes, no UNDO/DO pair can take search back to a previous state. So this space does not have the loops of the last, which appeared as chains of futile UNDO/DO actions. But it is still redundant, because the order in which fixes done matters. Suppose figure 11 is the problem. Then the top part of the search space is as in figure 12. States G and J are the same, though on different branches. (The move from B to F in an UNDO of x1.1.)

The redundancy can be limited by changing, not the space, but the strategy. The strategy is roughly depth-first, which means that it can be implemented with a stack, with corresponding savings of space, but it has restrictions on which arcs may be taken from any state.

--- Once a DO arc from some state has been taken, no other arc from that state can be taken on backtracking. (That is analogous to a Prolog cut.)

--- All fixes of a flaw must be rejected in a state before an UNDO arc can be taken from that state.

--- Once an UNDO has been applied to a state to remove a certain flaw, other UNDOs can be applied to that state on backtracking, but only UNDOs, and only if they remove the same flaw.

Different choices of which flaw to fix in any state on each run mean that each run may search a different subspace of the whole space. Each such subspace is nevertheless (I believe) complete. Part of one such subspace is figure 13. A is a failure state because w2 is a failing flaw. It can neither be fixed nor, since it is an original flaw, be cleared by UNDOing. So search in decision record space has to backtrack, ultimately to the alternative UNDO at U, and so reaches V. V is where search would have gone if the UNDO decision had been made differently. It was the inability to get there that made Hayes-like DDBT incomplete.

The only choice points that need be kept are those where UNDOs occur. If the and-or graph is a tree, so that each failing flaw can be undone in only one way, there will be no choice points, so the space can be searched by a deterministic process. That is why ordinary and-or tree search keeps a single set of decision records, (logically arranged in a tree though kept as a stack), rather than a whole space of sets of decision records.

Following only one DO out of any state may seem over rigid. But the choice of fix and flaw can either be quite arbitrary, or it can be made by imposing an order in which fixes and flaws should be tried. (Eg Prolog takes both flaws - goals in a clause - and fixes - clauses in a predicate - in the order in which they appear in the text.) The subspace in the example just shown corresponds to an ordering of flaws

$$w1 > w2 > w3 > w4$$

and an ordering of fixes

for w1	x1.1 > x1.2
for w2	x2.1 > x2.2

7.1   Is decision record space finite and complete? - disaster strikes

This space is finite if the and-or graph is finite. (Sketch proof: Take any branch in the state space tree. No flaw enters the branch more than once. The number of fixes either current or rejected for any flaw on the branch increases at each arc on the branch, but is limited by the size of the and-or graph.)

The space is not complete as it stands, because once a solution is found, there will be no failing flaw and so it will not be legal to apply an UNDO. But one needs to apply a forced undo to move on to the next solution. For instance, suppose the and-or graph of the problem is figure 14. Then the decision record space is figure 15.

And at this point it all goes wrong. The space is complete (I think) but redundant. What I previously proposed was an extra action that would drive one from one solution to another (for example, from A to B). Then all states of the dependency graph would be examined, and the space would be complete. The action was

FORCED UNDO If the state is a solution state, then strike through the   current fix of the "most urgent" flaw.

But in fact any obvious forced UNDO seems to have problems. Consider the very simple problem in figure 16. Search to the first solution is easy, but thereafter redundant, as in figure 17. Either both the forced UNDOs beneath solution 1 are allowed, so that solution 4 arises twice; or only one of them is allowed, in which case one only gets one of solutions 2 and 3.

And then there is another sort of redundancy that arises from a flaw being cleared twice as a result of only one failing flaw. The problem is shown in figure 18, and the corresponding space is figure 19, with its repeated states.

I have not been able to find any other way of going from one solution to the next without a similar problem arising. As a result, DDBT is unsolved.

8.   Where does the and-or graph come from?

Here is a separate issue that arises anyhow. Doing DDBT this way supposes that the and-or graph of the problem is available. The importance of having it is not so much as to be able to say what fixes are available for any flaw, as to be able to say which fix is a parent of which flaw.

One way of deriving the graph is by dividing the domain state into parts. Then a fix adds parts, and a set of parts constitutes a flaw. A fix is a parent of a flaw iff it adds a part which is part of the constitution of that flaw.

Deciding how to make the division of the domain into parts is one of the tasks of representing a domain. In planning, the kinds of parts I have found useful are: actions, orderings between actions, protected ranges and substitutions. The description of any fix is exhausted by giving the set of these it added, and the description of any flaw, by giving its type and the set of these that constitute it.

Figure 20 is a diagram, like a dependency graph, but showing how parts figure in it.

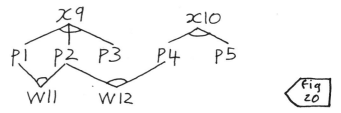

The corresponding dependency graph is figure 21.

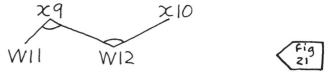

The only problem is that sometimes a part may be used in more than one fix. In planning, an action may have several effects, which may support the preconditions of several other actions. An example is figure 22.

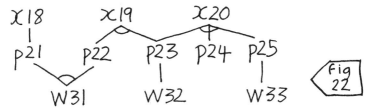

The dependency graph of this can be drawn if a distinction is made between and- and or- groupings of parents of flaws, as in figure 23.

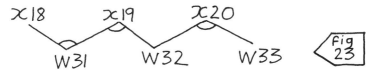

To get rid of a flaw with an or-group of parents (such as w32 above) each such parent must be undone. Doing this is just applying several UNDOs at once.

Acknowledgements

I thank Martin Henson, Richard Bartle and Jim Doran and members of the planning group at the University of Essex; and Austin Tate, of the Artificial Intelligence Applications Institute, University of Edinburgh. This work was partly supported by SERC grant GR/C/44938.

References

Bruynooghe M, Pereira LM: 1984
    Deduction revision by intelligent backtracking
    in: Implementations of Prolog, ed. Campbell JA. Ellis Horwood
Daniel L: 1977
    Planning: Modifying non-linear plans
    Working paper 24, Dept Artificial Intelligence, University of Edinburgh
de Kleer J: 1986
    An assumption-based TMS
    Artificial Intelligence 28(2) 1986
Dilger W, Janson A: 1986
    Intelligent backtracking in deduction systems by means of extended unification graphs
    J. automated reasoning 2(1) 1986
Doyle J: 1979
    A truth maintenance system
    Artificial Intelligence 12 (1979) 231-272
Hayes, Philip J: 1973
    Structuring of robot plans by successive refinement and decision dependency
    MPhil thesis, University of Edinburgh
Hayes, Philip J: 1975
    A representation for robot plans
    IJCAI-75 181-188
Levi G, Sirovich F: 1976
    Generalized and/or graphs
    Artificial Intelligence 7 (1976) 243-259

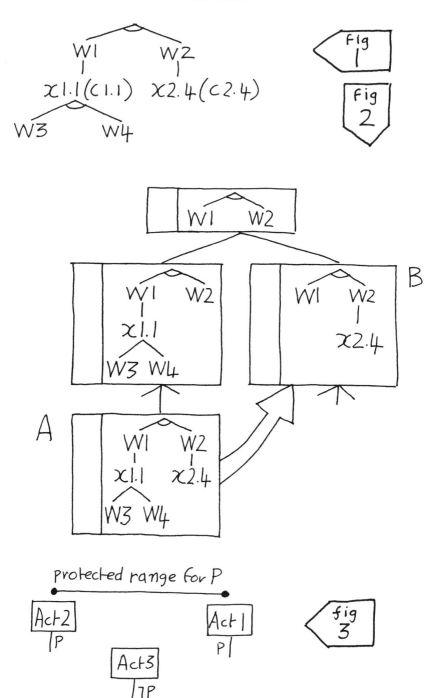

Fig
1

Fig
2

fig
3

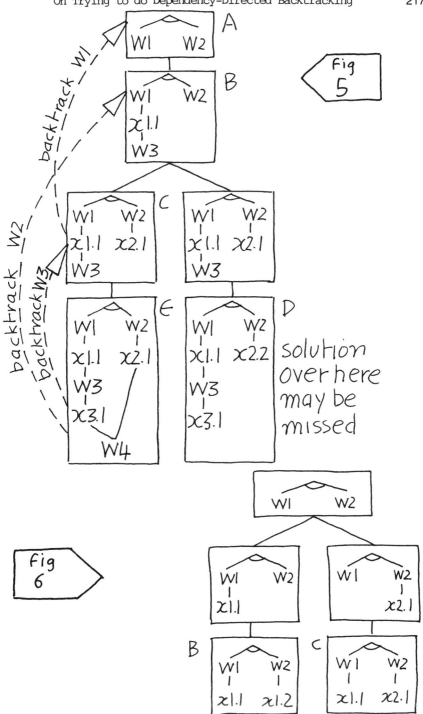

Fig
5

A
| W1   W2 |

B
| W1   W2
  x1.1
  W3 |

C
| W1   W2
  x1.1 x2.1
  W3 |

| W1   W2
  x1.1 x2.1
  W3 |

backtrack W1

backtrack W2

backtrack W3

E
| W1   W2
  x1.1 x2.1
  W3
  x3.1
        W4 |

D
| W1   W2
  x1.1 x2.2
  W3
  x3.1 |

solution
over here
may be
missed

Fig
6

| W1   W2 |

| W1   W2
  x1.1 |

| W1   W2
        x2.1 |

B
| W1   W2
  x1.1 x1.2 |

C
| W1   W2
  x1.1 x2.1 |

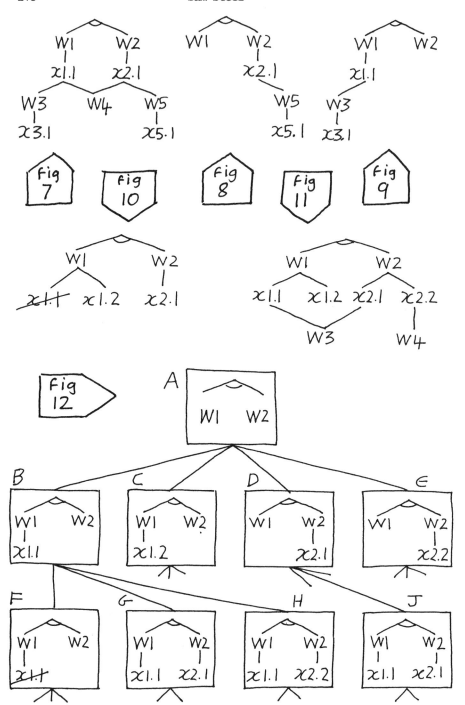

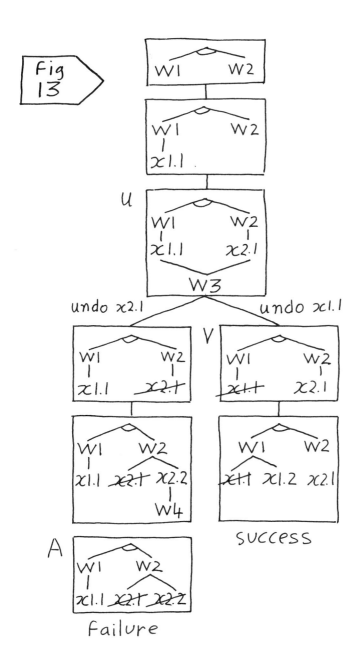

Fig 13

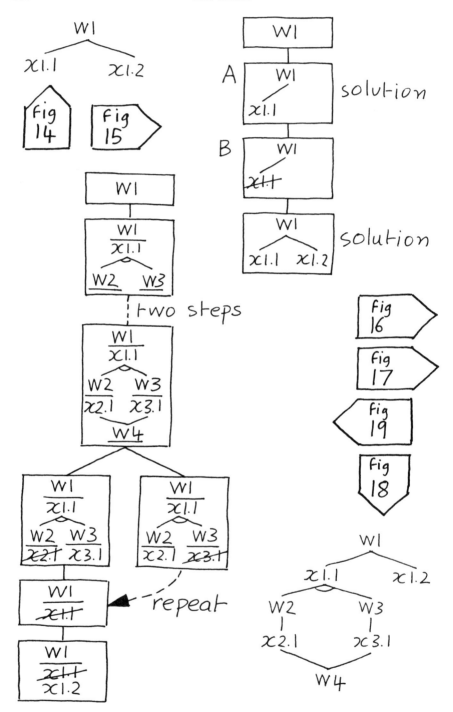

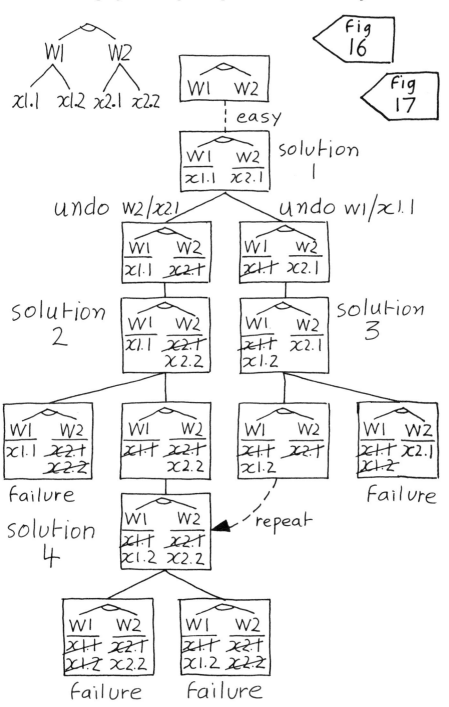

BIDIRECTIONAL CHART PARSING

Sam Steel, Anne De Roeck
Cognitive Science Centre / Dept Computer Science
University of Essex, Colchester CO4 3SQ, UK

ABSTRACT: This paper develops a suggestion of Thompson (1981) for an extension to chart parsing: that active arcs may be extended in either direction. Once this has been done, heuristics about which categories should be looked for in the string can be stated systematically and separately from the grammar proper. The paper describes a way of doing so.

1. Brief review of chart parsing.

Chart parsing is well described in (Thompson 1981). We review it using a variant notation, not because it is not known, but so that we have something to contrast our proposed extensions to.

1.1 The chart

Often a parsing process tries to analyze a string as a constituent of a certain category. While doing so, it recognizes the categories of substrings. But if it turns out the whole string is not of the category hoped for, so that the main parsing effort has to be abandoned, all the (perfectly valid) categories of the substrings get forgotten too. Chart parsing tries to avoid wasting that effort. The central idea is that all possible syntactic structures within a string are represented in a graph called the "chart" as soon as they are found. The nodes of the chart occur at the ends of the most basic parts of the string - typically, a word, perhaps a morpheme. For instance, in "The dog bit John", the nodes are disposed thus

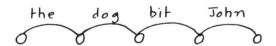

The arcs of the chart bear a label that indicates what the syntactic category of the substring between the start and finish of the arc are. Some arcs that give the category of some substrings of "The dog bit John" are

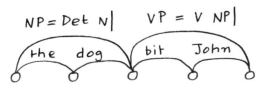

223

Examples of arcs that indicate what the category of substrings might be, if adjacent substrings are of certain categories are

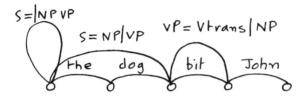

In general, the label of an arc is

    TotalCat = FoundCat... | UnfoundCat ...

This says

--- that there is a rule in the grammar

    TotalCat ::= FoundCat ... UnfoundCat ...

--- that the substring of the sentence between the ends of the arc contains an instance of FoundCat1, FoundCat2, ... in that order, and no substring that is not part of one of the FoundCats.

Either FoundCat ... or UnfoundCat ... may be null. Arcs that have no UnfoundCats in their label are said to be "inactive". All others are "active".

The categories FoundCat, UnfoundCat and so forth that occur on arcs are associated with substrings. One can also associate arbitrary structures (eg parse trees) with those same substrings. Then different analyses of a substring can be distinguished even if they are of the same category, and a chart parser is a true parser, not just a recognizer. However, the rest of the paper will talk as if arcs bear only categories. The provisos needed if extra structures are countenanced too are obvious.

1.2   The parsing process

A parse has been found when there is an inactive arc of the sought category from the beginning to the end of the entire string.

The parsing process is beautifully simple. When one has a meeting of arcs as here

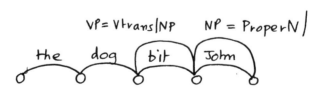

the active arc, which is looking for an NP, is allowed to extend itself by combining with the inactive arc, which represents a complete NP. As a result a new arc is added, to represent the fact that the combined strings constitute a complete VP.

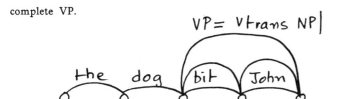

Though the new arc is here inactive, it is frequently active. No arc is ever deleted. No arc identical to one already in the chart is ever added.

If (1), (2),... are variables over nodes in the chart; X,Y,... are variables over categories of the grammar being used; <A>, <B>,... are variables over possibly null strings of categories; and <> is the null string of categories; then the rule for the event just described, called the "fundamental event" of chart parsing, can be stated as

Fundamental event rule: standard

```
(1)    X = ⟨A⟩ | Y ⟨B⟩    (2)
(2)    Y = ⟨C⟩ | ⟨ ⟩      (3)
---------------------------
(1)    X = ⟨A⟩ Y | ⟨B⟩    (3)
```

unless the resulting arc is already present.

1.3   Creating active arcs

The chart initially contains only inactive arcs, for the primitive elements of the sentence that were fed in. If the fundamental event is ever to take place, there must be some active arcs. Where do they come from?

There are two ways of getting them. An active arc corresponds to a hypothesis about what may be present in the string. Hypotheses can be formed top-down ("I want to find X - what would constitute an X?") or bottom-up ("I have found an X - what may it constitute?").

1.3.1   Top-down

In a grammar, top-down search amounts to saying

```
I seek a category X, starting at node (1)
There is a rule X ::= Cat1 ... in the grammar
So I will hypothesize a Cat1, starting at node (1)
```

If there are several rules re-writing X, then several arcs may be added. For example, the need to find a VP in

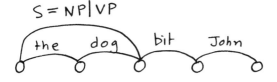

may lead to the top-down generation of the new arcs in

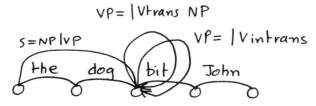

Formally, the top-down rule is

Adding active edges rule: top down: standard

```
(1)    X = ‹A› | Y ‹B›      (2)
Y ::= ‹C›
```

```
(2)    Y = ‹› | ‹C›         (2)
```

with the requirement that one start the process going by adding an active arc indicating what one is looking for in the whole string; if (begin) and (end) are the nodes they suggest, X is the category one is parsing for, and Dummy is not a category of the grammar, then the start rule is

Start rule: top down: standard

```
(begin)  Dummy = | X    (begin)
```

1.3.2   Bottom-up

The bottom-up strategy is

```
I have found a string of category X, starting at node (1
There is a rule Y ::= X ... in the grammar
So I will hypothesize a Y, starting at node (1)
```

If there are several rules re-writing to X ..., then several arcs may be added. For example, having found an NP in

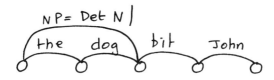

may lead to the bottom-up generation of the new arc in

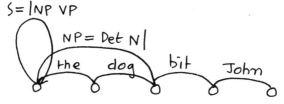

Formally, the bottom-up rule is

Adding active edges rule: bottom up: standard

```
(1)    X = ‹A› | ‹ ›           (2)
Y ::= X ‹C›
----------------------------
(1)    Y = ‹ › | X ‹C›         (1)
```

unless the resulting arc is already present.

There is no need for a start rule.

Issues about what rule one applies when matter for ease of use, but (except when there are categories that rewrite as the null string) as long as one applies all possible rules for the chosen strategy, all parses will be found.

2. Bidirectional chart parsing

2.1 Bidirectional chart parsing

That ends the review; what follows is our proposals.

There are rules that one wants applied in ways other than those provided above. One important case is the conjunction rule. In English, it is roughly true that if X is a category, then

X ::= X Conjunction X

is a rule. Suppose one is chart-parsing bottom-up. Then every time one finds an X, one has to add an active arc to see if that X is the first of a Conjunction. This is bad. Proposing active arcs is a lot of the cost of chart parsing. And each active arc added may be part of very many meetings of an active and inactive arc, each of which will have to be tested to see if it enables a fundamental event.

What one wants to do is to trigger the conjunction rule only once one has found a conjunction. But since the category Conjunction is not at the left end of the right-hand side of the rule in which it occurs, the bottom-up rule does not allow this.

Furthermore, if a conjunction did trigger the rule bottom-up, one would have to look for an X to the left of the conjunction one had found as well as, standardly, to the right.

The solution is to allow arcs in the chart to be incomplete on both sides. A label is then

```
TotalCat =
    LeftUnfoundCat ... | FoundCat ... | RightUnfoundCat ...
```

again remembering that any field may be null.

The bottom-up rule is now

Adding active edges rule: bottom up: bidirectional

```
(1)   X =  ‹ ›  |  ‹A›  |  ‹ ›   (2)
Y ::= ‹B› X ‹C›
-----------------------------------
(1)   Y = ‹B›  |  X  |  ‹C›    (2)
```

The top-down rule is going to come in two parts, depending which way one looks for an extension.

Adding active edges rule:
      top down: bidirectional: looking right

```
(1)   X = ‹L›  |  ‹F›  |  Y ‹R›   (2)
Y ::= ‹C›
-----------------------------------
(2)   Y = ‹ ›  |  ‹ ›  |  ‹C›       (2)
```

Adding active edges rule:
      top down: bidirectional: looking left

```
(1)   X = ‹L› Y  |  ‹F›  |  ‹R›   (2)
Y ::= ‹C›
-----------------------------------
(1)   Y = ‹C›  |  ‹ ›  |  ‹ ›       (1)
```

2.2 Bidirectional extension

There is now a problem with the fundamental event rule. An active arc may be extensible at both ends. These extensions will usually commute, so there will usually be two ways of producing the same arc. For example, suppose there is a configuration of arcs thus

```
(1)   A = ‹ ›|B|‹ ›   (2)
(2)   X = A|C|D     (3)
(3)   D = ‹ ›|E|‹ ›   (4)
```

Then there can be two fundamental events, yielding

```
(1)   X = ‹ ›|AC|D (3)
(2)   X = A|CD|‹ › (4)
```

But now each of these can combine with one of the original arcs, giving

```
(1)   X = ‹ ›|ACD|‹ ›  (4)
(1)   X = ‹ ›|ACD|‹ ›  (4)
```

If this is allowed, the chart will become redundant. There are three ways round this.

--- Allow this sort of commutation to take place, but check before any arc is added to make sure it isn't already there. This is simple, but it would be better if one didn't generate the redundant edges.

--- Demand that all unfinished ends of an arc be extended at once, by restating the fundamental event rule as

```
(1)    X = ‹›  |  ‹A›  |  ‹›        (2)
(2)    Y = ‹B›  X  |  ‹C›  |  Z  ‹D›   (3)
(3)    Z = ‹›  |  ‹E›  |  ‹›   (4)
------------------------------------
(1)    Y = ‹B›  |  X  ‹C›  Z  |  ‹D›    (4)
```

(which has two one-ended special cases; one is clearly the standard fundamental event rule). Then the only fundamental event available to the example set of arcs is to produce one copy of

```
(1)    X = ‹›|ACD|‹›  (4)
```

straight away.

This is elegant but elaborate. Each time one adds an arc to the chart, one will have to see if it completes any triple (rather than pair) of arcs that may enable a fundamental event.

--- Constrain extension. We insist that if an arc is extensible at both ends, it must extend rightwards. Only if it is inextensible to the right may it extend to the left. This amounts to choosing only one of the commuting routes. Obviously the other would have been equally good.

3. Heuristic information

But this machinery is pointless unless one indicates whether a rule is to be used top-down or bottom-up, and, if bottom-up, which categories are to trigger it. The simplest way to do this is to make all use in a run either top-down or bottom-up, and to make the first category on the right the invariable trigger. But that is too crude - as with the conjunction rule. We want to be able to give the heuristics

--- Systematically, by rule rather than by list.

--- Separately from the grammar, which should be about language, not parsing.

--- Locally. Some rules are best used up, some down. That decision should be local to the rule.

We propose the use of the annotations {up} and {down} to do this. These restrict the adding of new active edges, so that only appropriately annotated grammar rules can be used. They do not affect the grammar at all. Their meaning is exhausted by saying how they affect the parsing process. That is done by giving the revised formal rules for adding active arcs:

Adding active edges rule:
bottom-up: bidirectional: with annotation

```
(1)    X =  ‹›  |  ‹A›  |  ‹›   (2)
Y ::= ‹B›  {up}X  ‹C›
-----------------------------
(1)    Y = ‹B›  |  X  |  ‹C›    (2)
```

```
Adding active edges rule:
     top-down: bidirectional: with annotation
(1)    X = ‹A› | ‹B› | Y ‹C›    (2)
{down}Y ::= ‹D›
--------------------------------
(2)    Y = ‹ › | ‹ › | ‹D›        (2)
```

(There is a similar top-down rule for when the Y is to be found on the left; then the added edge will look for <D> to the left too). (Similar annotations for the standard rules are obvious.)

All annotations are valid in every run, so some rules will be used top-down while others are used bottom-up. A rule may bear more than one annotation, so it may be used several ways, though we have no examples of this being useful.

3.1 Benefits working bottom-up

The conjunction rule, used to motivate bidirectional parsing, should now be stated as

X ::= X {up}Conjunction X

X is a variable over categories, which must be bound everywhere alike, in order to prevent "*The cat and John walked" being parsed as a conjunction. Allowing variables over categories onto arcs presents only trivial problems.

3.2 Benefits working top-down: categories that rewrite as null

There is of course a problem in looking bottom-up for a category (say N) that rewrites as null. There will be nothing in the input string which counts as a reason to look for N. Since N is not found, rules in the grammar that have N on the right will not be considered, so the categories on those rules' left sides will not be proposed, and so on. And so a parse of the string may be missed.

One high-minded but impractical strategy would be to accept that a grammar will contain rules of the form

X/X ::= ‹ ›

for all X such that X is a category of the grammar. Since there is no way of telling in advance where traces lie, one must hypothesize them everywhere. So each node of any chart is seen as having inactive arcs for every sort of trace running from itself to itself. The rule licensing such arcs' presence is

```
X ::=   ‹A›
--------------------------------
(1)    X/X = ‹ › | ‹ › | ‹ ›        (1)
```

But this generates far, far more arcs than is reasonable.

Another, better, solution to this is to attach more heuristic information to the grammar rules. One wants to be able to say that some rules may be used to look for a category top-down. This is the function of the annotation {down}, which can be used as here:

```
X ::= ‹A› Y ‹B›
    =›
X/Z ::= ‹A› Y/Z ‹B›
```

Then the rule

```
S ::= NP VP
```

will generate, among others,

```
S/NP ::= NP/NP VP
S/NP ::= NP VP/NP
```

But if the initial rule bore the heuristic information

```
S ::= NP {up}VP
```

should this be carried over to the generated rules, to make them

```
S/NP ::= NP/NP {up}VP
S/NP ::= NP {up}VP/NP
```

? No. It would be foolish to look for a trace NP to the left of any VP found. Nor can the meta-rule be trusted to insert the annotation. In general, if there is a rule of the form

```
X ::= ‹A› {up}Y ‹B›
```

one will want to annotate its slashed form so as to give

```
{down}X/Z ::= ‹A'› Y ‹B'›
```

(where the primes on A', B' just indicate that a /Z occurs somewhere) and one might find ways to do that. But suppose one had a rule with a distinctive (perhaps syncategorematic) item in it, something such as

```
X ::= ‹A› {up}distinctive_item ‹B›
```

Any such process would probably take that rule and give

```
{down}X/Z ::= ‹A'› distinctive_item ‹B'›
```

whereas one wants to continue to use the rule bottom-up and triggered by the distinctive item in it:

```
X/Z ::= ‹A'› {up}distinctive_item ‹B'›
```

Our solution is to attach annotation by rule. That is, there is one component of the system that generates rules of the grammar, and another that annotates them. The rules of annotation that evade the problem just pointed out can be stated as

```
if there is a rule
        VP ::= Vtrans NP
then there is an annotated rule
        VP ::= {up}Vtrans NP

if there is a rule
        X ::= ‹A›
and ‹A› = ‹A1› distinctive_item ‹A2›
then there is an annotated rule
        X ::= ‹A1› {up}distinctive_item ‹A2›
```

```
VP/NP ::= {up}Vtrans NP/NP
{down}X/X ::= < >
```

So if there is an arc

(1)   VP/NP = < > | Vtrans | NP/NP < >   (2)

in the chart, then we can legitimately apply the top-down rule, and add an arc

(2)   NP/NP = < > | < > | < >   (2)

So the null string has been sought and (simultaneously) found.

3.3   How to add annotations

The rules for relative clauses persuaded us that it was rules that must bear annotations, not categories. We used an intentionally crude GPSG-like grammar (Gazdar et al 1985) with the rules

```
N ::= N RelClause
RelClause ::= that S/NP        unreduced relative clause
RelClause ::= NP VP/NP         reduced relative clause
```

to cover the well-known facts

    The man that Celia met died.
    The man that met Celia died.
    The man Celia met died.
*   The man met Celia died.

To exploit a "that", while still being able to find reduced relatives, one has to annotate the rules as

```
N ::= {up}N RelClause
RelClause ::= {up}that S/NP
{down}RelClause ::= NP VP/NP
```

The merit of this is that, when a noun is found, reduced relative clauses have to be looked for, but unreduced ones don't. If the category RelClause bore an annotation valid in all rules, it would have to be {down}, and unreduced relatives would be sought even when there was no "that".

At first we attached the annotation directly to the rules of the grammar. Going bottom-up, only one category would be marked {up}. Consider

```
NP ::= Det N
```

There is no point in both Det and N triggering the rule. After all, it will only succeed if both are present, so either can be the key without causing incompleteness. As a rule of thumb, we annotated the category that occurred in the right-hand side of the fewest rules. The ideal is to have a syncategorematic terminal, such as the "that" in a relative clause.

But annotation can cause trouble when rules can be generated by meta-rules (Gazdar et al 1985), which are devices that say "Whenever there is a rule of this (given) form in the grammar, then there is also another rule of this (related) form in it". For instance, one might state the meta-rule that introduced slash categories (Gazdar et al 1985) as

```
X  ::=  ⟨A⟩  Y  ⟨B⟩
       =⟩
X/Z  ::=  ⟨A⟩  Y/Z  ⟨B⟩
```

Then the rule

```
S  ::=  NP  VP
```

will generate, among others,

```
S/NP  ::=  NP/NP  VP
S/NP  ::=  NP  VP/NP
```

But if the initial rule bore the heuristic information

```
S  ::=  NP  {up}VP
```

should this be carried over to the generated rules, to make them

```
S/NP  ::=  NP/NP  {up}VP
S/NP  ::=  NP  {up}VP/NP
```

? No. It would be foolish to look for a trace NP to the left of any VP found. Nor can the meta-rule be trusted to insert the annotation. In general, if there is a rule of the form

```
X  ::=  ⟨A⟩  {up}Y  ⟨B⟩
```

one will want to annotate its slashed form so as to give

```
{down}X/Z  ::=  ⟨A'⟩  Y  ⟨B'⟩
```

(where the primes on A', B' just indicate that a /Z occurs somewhere) and one might find ways to do that. But suppose one had a rule with a distinctive (perhaps syncategorematic) item in it, something such as

```
X  ::=  ⟨A⟩  {up}distinctive_item  ⟨B⟩
```

Any such process would probably take that rule and give

```
{down}X/Z  ::=  ⟨A'⟩  distinctive_item  ⟨B'⟩
```

whereas one wants to continue to use the rule bottom-up and triggered by the distinctive item in it:

```
X/Z  ::=  ⟨A'⟩  {up}distinctive_item  ⟨B'⟩
```

Our solution is to attach annotation by rule. That is, there is one component of the system that generates rules of the grammar, and another that annotates them. The rules of annotation that evade the problem just pointed out can be stated as

```
if there is a rule
        VP  ::=  Vtrans  NP
then there is an annotated rule
        VP  ::=  {up}Vtrans  NP

if there is a rule
        X  ::=  ⟨A⟩
and ⟨A⟩  =  ⟨A1⟩  distinctive_item  ⟨A2⟩
then there is an annotated rule
        X  ::=  ⟨A1⟩  {up}distinctive_item  ⟨A2⟩
```

```
if there is a rule
        X/Y ::= ‹A›
and ‹A› =/= ‹A1› distinctive_item ‹A2›
then there is an annotated rule
        {down}X/Y ::= ‹A'›
```

If one wanted to change the parsing process to pure top-down, all one would need to do would be ensure that the start rule was present, and have the single annotation rule

```
if there is a rule
        X ::= ‹A›
then there is an annotated rule
        {down}X ::= ‹A›
```

To do the sensible thing with words, and spot that "dog" is a noun on input, rather than looking for all nouns as part of finding N top-down, one should add the annotation rule

```
if there is a rule
        X ::= Y
and Y is lexical
then there is an annotated rule
        X ::= {up}Y
```

To work pure bottom-up, one would need just the annotation rule

```
if there is a rule
        X ::= Y ‹A›
then there is an annotated rule
        X ::= {up}Y ‹A›
```

There is a price to be paid for having more control over the use of rules, since adding the wrong control may lose one parses. Suppose one has a grammar in which the only rules mentioning category Y are

```
X ::= A Y B
Y ::= J K
```

and one annotates them as

```
X ::= A {up}Y B
{down}Y ::= J K
```

then neither will ever be executed. It is not clear to us what the general conditions are, either for falling in to this trap, or for avoiding it. Nevertheless, we have found the added control well worth having.

4. Related work

Bidirectional extension of active edges has also been implemented by du Boulay and Elsom-Cook (1986). Their interest is in detecting and advising on the correction of errors in program code. They opportunistically identify "islands of reliability" in the input code, and try to extend it in each direction. Their parser permits mixed top-down and bottom-up creation of active edges. Since their purposes are different from ours, they do not consider parsing natural language, nor do they describe the details of bidirectional extension, nor how it interacts with the phenomena we have described.

5. Conclusions

Bidirectional chart parsing is an easy extension to ordinary chart parsing. Heuristic information can be added so that bidirectionality is used only when it is useful. We have implemented such a system and used it in a natural language front-end project. Though we present no figures (to make them meaningful is a hard task), we are very pleased by the improvement in speed and space that these improvements produce.

Acknowledgements

The ideas on bidirectional parsing in this paper owe much to Henry Thompson, for whom Sam Steel formerly worked as a research assistant. Anne De Roeck discovered the need for it independently. She was supported by a grant from ICL. Our thanks to our referee for his incisive comments.

References

du Boulay B, Elsom-Cook M: 1986
        A PASCAL program checker
        ECAI-86

Gazdar G, Klein E, Pullum G, Sag I: 1985
        Generalized phrase structure grammar
        Blackwell

Thompson H: 1981
        Chart parsing and rule schemata in PSG
        19th annual meeting of Assoc. Computational Linguistics

Human Problem
Solving and
Programming

TRANSFER OF LEARNING IN INFERENCE PROBLEMS: LEARNING TO
PROGRAM RECURSIVE FUNCTIONS.

M.A. Conway,
MRC Applied Psychology Unit, Cambridge.

H. Kahney,
Pscychology Dept., The Open University, Milton Keynes.

ABSTRACT

Two studies investigated student programmers solving sequences of
analog programming problems involving recursion. In the first study
students provided protocols while solving on-line problems.
Analysis of the protocols indicated that students employed
inappropriate mapping strategies from example problem and solution
to current problems. Furthermore, students showed little evidence
of understanding the abstract relations underlying sequencies of
problems. Student's mappings were at a surface level. In a second
study, separate groups of students were provided with training
problems, solutions, and additional information. Students provided
with a formal definition of recursion and an appropriate mapping
relation showed significantly more transfer of learning than other
groups. These findings indicate that current models of program
problem solving (e.g. Anderson, Farrell, & Saunders, 1984) may be
too powerful to accurately characterise the learning process.
Moreover, text books and training manuals rarely specify abstract
and mapping relations. The present findings indicate that when this
information is explicitly stated learning is facilitated.

INTRODUCTION

One way in which students learn to program is by applying the
outcomes of previous learning episodes to a current problem. Such
transfer of learning has, however, proved particuarly difficult to
demonstrate in the laboratory (e.g. Gick & Holyoak, 1980; Kahney,
1982; Reed, Dempster, & Ettinger, 1985; Pirolli & Anderson, 1985)
and it appears that problem solvers are unlikely to perceive
relations between similar or analog problems unless specifically
instructed. Furthermore, successful transfer of learning,
especially in solving successive programming problems, may be far
more complex than simply detecting a similarity between two problems
(Holyoak, 1985). The present research, then, examines in detail how
student programmers solve successive analog problems, how naturally
occurring problem solving strategies may impair transfer of
learning, and how transfer of learning may be facilitated.

STUDY 1: Solving Analog Recursion Programming Problems

In this study we collected protocols from student programmers while they solved sequences of similar problems. All the problems involved writing a short programme which contained a recursive function. Pirolli & Anderson (1985) report a similar series of studies in which three subjects provided protocols of attempts to solve recursive programming problems. The central finding of Anderson & Pirolli was that their student programmers repeatedly attempted to map a previous problem solution onto the current problem statement. Later success in problem solving was attributed to students developing abstract representations of the recursive function. The present study examines these aspects of analogical problem solving in further detail.

The subjects and language

The subjects were six Open University cognitive psychology students who had completed a course in which they learnt the symbolic processing language 'Solo' (Eisenstadt, 1978). Solo is a LOGO-like database manipulation language.

The recursion problems

All students were presented with a prototypical problem and solution couched in everyday terms in order to simply the explanation of recursion: "Imagine a chain of 'KISSES' relations, e.g. JOHN KISSES MARY KISSES FRED KISSES JANE, etc. A procedure called INFECT can propagate FLU all the way through the chain of KISSES relations, so we end up with JOHN HAS FLU, MARY HAS FLU, etc." The example is explained to the students in great detail, including several pages of text, diagrams, and a worked-through trace of a sample invocation of INFECT. The analog problems are formally identical both in problem structure and solution to the example problem. Examples of analog problems are described below. In all cases the student's task was to write a procedure which would propagate the influence. Note that all students had previously written a formally identical procedure as part of their course work. Students were instructed and prompted to 'think aloud' while problem solving.

An interpretation theory for analyzing protocols

The protocols provided by novices pose particularly difficult problems of interpretation because they often reflect only a very poor understanding of the domain of learning and because they are often highly idiosyncratic. In order to analyse the protocols an interpretation theory was developed (c.f. Kahney, 1982; 1986).

In our view the problem solving process is based on two general apsects of student's problem solving behaviour: Information Gathering and Mapping Processes. Interpretation of the Mapping Process is based on theories of problem solving proportional analogies. Proportional analogies are analogies of the form:

        A : B :: C : ?
        e.g. Letter is to word as sentence is to ?WHAT?
Solving such analogies involves a number of component processes such
as understanding the different terms of the analogy (leter, word,
sentence), inferring the mapping between the A and B terms,
inferring the mapping between the A and C terms, and applying the A-
B mapping to the C term in order to derive the D term, or the answer
to the analogy. In our interpretation theory for analogical problem
solving the training problem presented in the programming manual
(the INFECT problem) is represented as the A term of the
proportional analogy. The solution to the training problem is
represented as the B term. The current exercise problem is
represented as the C term, and the solution to be achieved by the
student is, of course, the ?D term of analogy.

Table 1 summarizes a selection of rules for classifying protocol
statements as an episode involving not only one of the mappings
outlined above other possible (but somewhat illogical) mappings.

        Table 1: Six possible mappings.

MAPPING	Example/Explanation
1 TRAING–PROBLEM–GIVENS mapped onto TRAINING–PROBLEM–SOLUTION	Determine relation between training problem in textbook and its solution.
2 EXERCISE–PROBLEM–GIVENS mapped onto TRAINING–PROBLEM–GIVENS	Determine relation between exercise problem and the training problem
3 TRAINING–PROBLEM–GIVENS mapped onto EXERCISE–PROBLEM–SOLUTION	Determine relation between the training problem and the solution to the exercise problem.
4 TRAINING–PROBLEM–SOLUTION mapped onto EXERCISE–PROBLEM–GIVENS	Determine relation between training problem solution and the exercise problem
5 EXERCISE–PROBLEM–SOLUTION mapped onto TRAINING–PROBLEM–SOLUTION	Determine relation between exercise problem solution and the training problem solution
6 EXERCISE–PROBLEM–GIVENS mapped onto EXERCISE–PROBLEM–SOLUTION	Determine relation between exercise problem and its solution

The protocol interpretation theory proposed here also posits Problem
Understanding and Problem Solving processes (after Hayes & Simon,
1974). The Problem Understanding process occurs not only when a
student is first confronted by a problem statement but continually
in protocols, often following a fairly brief Problem Solving
episode. Problem Solving episodes often provide new information or
understanding (or new misconceptions or misunderstandings) which the
student uses in order to reformulate the given problem. For a full
account of this aspect of the interpretation theory see Kahney
(1982; 1986).

Results and Discussion

In order to test their understanding of recursion, we investigated students solving several recursive inference problems, including one based on a real world example that, on the basis of past work (Kahney, 1982), we knew to be a problem our students could readily comprehend. The problem is as follows:

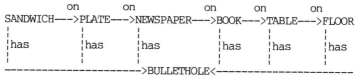

```
          on          on          on          on          on
SANDWICH--->PLATE--->NEWSPAPER--->BOOK--->TABLE--->FLOOR
```

Write a program which simulates the effect of someone firing a very powerful pistol aimed downwards at the topmost object (SANDWICH), yielding the final database shown below:

```
          on          on          on          on          on
SANDWICH--->PLATE--->NEWSPAPER--->BOOK--->TABLE--->FLOOR
    |           |          |         |        |        |
    |has        |has       |has      |has     |has     |has
    |           |          |         |        |        |
    ------------------------>BULLETHOLE<---------------------
```

The SHOOTUP problem is an interesting test case of a student's ability to solve problems by analogy because (1) the mapping between the A-B terms of the analogy (the INFECT problem and its solution) have already been precisely elaborated in the details of the worked out example in the programming manual, (2) the SHOOTUP problem is completely isomorphic to the training problem, a fact which should facilitate transfer, and (3) the problem statement for the SHOOTUP problem points the solver, in the case of difficulties, to the discussion of recursion in the programming manual in which the INFECT problem features. This hint to use the INFECT problem as a framework for solving the SHOOTUP problem obviates the need for the novice to work out for him/herself which example program should be imitated. On the other hand, the SHOOTUP problem statement does not make explicit the exact nature of the relationship between the INFECT and SHOOTUP problems. In fact, the two problems seem very different on the surface, the one dealing with the spread of 'flu' through a social act, the other involving a senseless act of destruction. In order to determine the isomorphism between the two problems, the novice must construct a more abstract problem representation. This involves understanding that in SOLO tail recursive programs are supported by a chain of nodes connected by a single relaton (such as KISSES or ON) in the database. The SHOOTUP problem presents the novice with an opportunity to compare and contrast two seemingly unrelated problems (surface structure) at an abstract level, and hence to acquire a schema for problems in recursion.

We presented the SHOOTUP problem to the six students after they had worked their way through the SOLO programming manual and worked all of the exercise problems. Of the six, four resorted to imitation of the INFECT program as their main solution method. In order to illustrate this we present the following excerpts from subjects S3's protocol. S3 typifies the behaviour of our subjects. S3 eventually solved the SHOOTUP problem by using the INFECT program as an analog solution framework.

Note that statements made by S3 are numbered, S3's first statement
was numbered one. Problem solving episodes are titled according to
the interpretation outlined above and in Kahney (1982; 1986). 'E'
refers to statements made by the experimenter.

MAPPING CHECK-primitive -> SHOOTUP-GIVENS
72 But, er, knowing the constituents of normal sandwiches, normal
plates, normal newspapers, books, and so on and so forth, I accept
that there is going to be a hole going right through.
73 It is for this reason that if I go for CHECK procedure and if I
do this, yes, fine, 'If Present: CONTINUE'; If Absent', well, that's
where I get into a little bit of a problem, a turmoil in my own mind
as much as that pattern, although it arises in the [CHECK]
procedure, doesn't happen in this problem here.
E: What pattern are you looking for?
74 A yes or no pattern.
75 CHECK if bullet goes through sandwich in order to follow the
pattern down.
76 Um, in this case I'm prompted to justify this 'no' by saying in
effect to, to say, well, question, question your answer because this
is, your answer is questionable and so loop it back to the beginning
again and start again. In other words if you always come up with a
questionable answer you're going to work on a virtual continous loop
and this is going to be a problem, uh, for the simple reason....

UNDERSTANDING SHOOTUP-GIVENS
78 I'm reading the problem again, I'm looking down and seeing all
the 'has, has, has's' in case I've missed anything out because it
seems so simple on the surface of it.
79 I'm probably trying to find, trying to read something into it
that's not there.
E: Do you have any idea what that might be?
80 Well this is the reason why I went back to the problem and saw if
there were any 'not's' in there that I'd overlooked in the first
place because I was not wanting to see 'not's'.
E: You're re-reading the problem?
81 Yes, I'm looking up here and seeing the "The database represented
by the state of affairs initially described..."

MAPPING INFECT-SOLUTION -> SHOOTUP-GIVENS
86 Um, I reckon the first INFECT procedure that I can see on this
page and trying to implant this SHOOTUP problem onto this INFECT
procedure.
87 If we say 'SHOOTUP X'.
88 Umm, one, NOTE...
89 I'm looking at the [SHOOTUP] problem again.
90 X HAS BULLETHOLE.
91 Um, looking again to 'CHECK' [at step 2 of the INFECT program].
92 Well, that doesn't apply.
E: That doesn't apply?
93 Yes.

UNDERSTANDING SHOOTUP-SOLUTION

94 If we..., X is the variable...so if we can replace X by either,
umm, the sandwich, the plate, the newspaper, the book, the table,
the floor.
95 So, TO SHOOTUP SANDWICH.
96 NOTE SANDWICH HAS BULLETHOLE.
97 NOTE PLATE HAS BULLETHOLE.
98 NEWSPAPER HAS BULLETHOLE, and so on and so forth.
99 So if I put that into SOLO, SOLO would come back and say, OK, he
knows how to SHOOTUP.
100 If I do that then I'm thinking that if I put 'FIDO', then FIDO
will have a bullethole and FIDO is not necessarily included in this
lot here and the SHOOTUP procedure is only applicable to these six
items.

MAPPING INFECT-SOLUTION -> SHOOTUP-SOLUTION

E: What are you reading?
106 I'm reading, I'm scanning this procedure here, the INFECT
procedure here again because that, I think that this procedure [the
reference here is to the subject's own current solution to the
shootup problem], leastwise NOTE X HAS BULLETHOLE will work.
107 Uh, but it will work overly, it'll work for everything which is
slotted into X and we don't want everything, we only want six items.

MAPPING INFECT-GIVENS -> SHOOTUP-GIVENS

111 I'm looking at the diagram here: "LIZ KISSES JOE. LIS HAS FLU,
JOE HAS FLU.
112 JOE HAS FLU is the inference.
113 If I replace that and say GUN SHOOTS SANDWICH...
114 No, that won't work.
115 That inference won't work.
116 Let's try to get someting down on paper.
117 Let's try to get something down on paper that makes the
inference.
118 We've got a gun. Shoots, shoots bullet, gun is on sandwich.
119 We're inferring that the bullet goes through the sandwich.
120 Gun shoots bullet, gun is on sandwich, bullet through sandwich.
121 So that's exactly the same as they've got there, page 80.

In the first episode of the protocol presented above, S3 attempted
to map the primitive CHECK procedure (SOLO's 'If-Then-Else
construct) onto the SHOOTUP-GIVENS. Unfortunately, the subject had
a serious misunderstanding about the CHECK procedure, which has the
following structure:
CHECK <node relation node>
If Present: <action-1>; CONTINUE or EXIT
If Absent: <action-2>; CONTINUE or EXIT
S3 thought that an action of some sort was obligatory on the 'If
Present' and 'If Absent' branches of the CHECK statement, rather
than optional, and as a result spent a lot of his time trying to
find something in the SHOOTUP problem statement that indicated what
to do in the case of a pattern being absent from the database. The
subject eventually concluded that the CHECK statement was irrelevant
to the problem (lines 91-93), a very serious mistake.

An extremely interesting feature of the protocol is S3's attempt to map the relation between the INFECT problem givens and the SHOOTUP problem givens. The mapping occurs at a surface level, that is, at the level of the objects mentioned in either problem statement (lines 111-121). Also, even though S3 pronounces a successful mapping (line 121), it can be shown that the mapping even at the surface level is incorrect. This is interesting because Gick and Holyoak (1980) have suggested that people are very good at solving problems by analogy once they are given a hint that two problems are related.

The behaviour of this subject, and of our other three subjects who attempted to solve the SHOOTUP problem by analogy to the INFECT problem, indicate that subjects are very poor at making such mappings. We believe that the discrepancy between our findings and those of Gick and Holyoak can be explained in terms of a contrast between the relatively simple type of analogy problem which Gick and Holyoak used in their research, involving only one step in the solution process, and the greater complexity of programming problems, which involve a number of steps to achieve a solution. Our analyses suggest that the relationship between a training problem and exercise problems should not be left to students to puzzle out for themselves, but should be made explicit. Although this might seem just plain common sense, the fact is that many textbook writers do not employ such principles.

Using the interpretation theory outlined above, we segmented S3's protocol into 21 main episodes, an extract of which is shown below in Table 2. In Table 2, groups of episodes have been clustered together to reveal patterns in mapping episodes.

Table 2: Problem solving episodes

I.
1 INFORMATION GATHERING: *CHECK
2 MAPPING CHECK --> SHOOTUP-GIVENS
3 UNDERSTANDING SHOOTUP-GIVENS
4 MAPPING CHECK-primitive -> SHOOTUP-GIVENS
5 UNDERSTANDING SHOOTUP-GIVENS
6 INFORMATION GATHERING: *INFERENCE
II.
16 MAPPING CHECK -> SHOOTUP-GIVENS
17 INFORMATION GATHERING: *WILD-CARD-PATTERN-MATCHING
V.
18 MAPPING WILD-CARD-PATTERN-MATCHING -> SUSS-SOLUTION
19 UNDERSTANDING INFECT-SOLUTION
VI.
20 MAPPING INFECT -> SHOOTUP-GIVENS
21 MAPPING SHOOTUP-SOLUTION -> SHOOTUP-GIVENS

These analyses reveal that most students possess very little skill at solving problems by imitation of worked out examples, and learn very little from such examples. For one thing, they are unable to infer the deep or abstract relationship between isomorphic problems, mainly because they never go beyond comparisons of problems at the surface level. We also find that novices rarely attempt to map the statement of a training problem onto the solution but rather prefer to begin work on an exercise problem by trying to map the solution to the training problem directly onto the givens of their exercise problem. This is equivalent to trying to solve the analogy 'Letter is to word as sentence is to ?WHAT' by mapping 'word' onto 'sentence', and ignoring the term 'letter' altogether. In other words, students naturally occurring problem solving strategies entail the use of inappropriate mappings between successive problems. It is thus not surprising that novices often fail to solve problems by analogy. Moreover, these findings suggest that the analogical problem solving mechanisms built into machine learning programs such as ACT (Anderson, 1983) are far too powerful to serve as accurate models of human analogical problem solving processes.

STUDY 2: Transfer of learning in solving recursion problems

In this study we examine transfer of learning in large groups of student programmers, similar to those used in the previous study. Different groups of students are provided with either A) the INFECT problem and solution (Solution only group), B) The INFECT problem and solution plus a formal definition of recursion (Definition group), C) The INFECT problem and solution plus a mapping relation identifying the types of SOLO databases which support recursion (Mapping relation group), D), The INFECT problem and solution plus both the formal definition of recursion and the maping relation (Definition and Mapping group). Students first study the INFECT problem and attempt to solve it. Students are then provided with the solution, plus additional material depending on group, and instructed to compare their own solution with the one provided.

After this initial training stage students then attempt to solve three successive problems. One problem was a close isomorph of INFECT (Close problem), a second problem was a distant isomorph (Distant problem), and a third problem was also a distant isomorph which, unlike the other two test problems, was not accompanied by a database. In this third problem students must solve the problem and generate their own database (Database problem).

The experiment is very similar in design to experiments reported by Reed, Dempster, and Ettinger (1985) and Pirolli and Anderson (p.270, 1985). On the basis of our protocol study we expected that the Definition and Mapping group should show the most marked transfer of learning. This is because the isomorphic nature of the problems are explicitly pointed out to them and the problems are formally defined as problems in recursion. Compared to the Solution only group we expected that the Definition only group and Mapping only group would show elevated levels of transfer. It was not clear, however, whether the Definition group would differ from the Mapping group. It could be argued that because the Definition group receive the

abstract information required to solve the problems then they should
show highest levels of transfer (Pirolli & Anderson, 1985).
However, the results of our protocol study indicated that because
students attempt inappropriate mappings from training problems to
test problems then the Mapping group may show relatively high levels
of transfer.

In addition to these predictions we expected that, in general, Close
problems should show more transfer than distant problems (Reed et
al., 1985) and that the Database problem should show least transfer.
Although the Maping only and Definition plus Mapping groups may show
transfer to this problem because the correct type of database has
been identified for them.

Methodological details

A total of 180 SOLO student programmers took part. All were mid-way
through the SOLO course and had similar programming experience to
the students in the protocol study. Students with other types of
programming experience were excluded from the sample. The students
undertook the experiment as part of a residential summer school and
were paid for their participation. There were 57 students in the
Solution only group, 43 in the Definition only group, 34 in the
Mapping only group, and 46 in the Definition plus Mapping group.

Students were tested in four groups in a large lecture theatre on
different days. In the training phase of the experiment students
were allowed five minutes to study the example problem, 10 minutes
to write a program solution and five minutes to study the solution
provided. The students retained these materials throughout the
experiment and were encouraged to refer back to them. Students then
solved the test problems and were allowed 10 minutes for each
problem. After attempting a solution, the problem and attempted
solution were removed. Order of presentation of problems was
counterbalanced within each group, however, as the order in which
students solved problems did not produce any reliable effects this
factor is not discussed further. A single testing session lasted
about one hour.

Results and Discussion

The mean percentage of correctly solved problems in each group is
shown in Table 3, below.

Table 3: Mean percentage of correct solutions

		Problem type		
	Training	Close	Distant	Database
Groups				
Solution only	5.3	17.3	24.5	21.0

Definition only	O	27.0	18.0	18.0
Mapping only	O	32.0	29.0	5.0
Definition plus Mapping	4.0	52.0	47.0	29.0

Solutions were classified as correct if it was judged that the program would function as specified when compiled. Incomplete syntax and, for working solutions, ordering of code were ignored. Programs were scored by two independent judges and close inter-judge agreement was observed, .87. Disputed solutions were entered in the class 'Other' in an analysis of errors. This analysis is, however, not reported in detail. The data was analysed with Chi-square tests and for all Chi's reported $p < 0.05$.

All groups showed a significant effect of transfer of learning (chi=67.23). However, not all groups differed significantly in the extent of transfer of learning. Only the Definition plus Mapping group produced significantly more correct solutions than the control Solution only group (chi=23.7). All three problem types were reliably solved more frequently by students in the Definition plus Mapping group than by students in any of the other groups. In this group there was no reliable difference between the number of Close and Distant problems correctly solved, but significantly less Database problems were solved in comparison to the other two types of problem (chi=18.1).

Overall, no differences were observed between the Definition and Mapping only groups. The Mapping only group solved significantly more Distant problems than the Definition group (chi=15.3), however, the Definition group solved significantly more Database problems than the Mapping only group (chi=20.1). These findings are difficult to interpret because the Solution only group solved more Database problems than either group. A problem here relates to students specification of their databases which counted as part of the solution. It seemed that some students felt it sufficient to specify only part of the database as an example of the type of database. Unfortunately, it was not possible to verify this with students and we can only conclude that interpretation of solutions to the Database problem must await further research.

Finally, a classification of error programs was undertaken. This will not be reported in detail here, however, one striking finding should be mentioned. Erroneous solutions in the Definition plus Mapping group, for all problem types, were characterised by the incorrect or incomplete use of 'self' statements. 'Self' statements comprised the recursive call in which the program calls itself. The other groups rarely showed this error and either omitted self statements altogether or (generally in the few cases of correct solutions) contained correct self statements. The predominant error in these groups related to the incorrect specification of conditionals used to search the database. Note that this was also a problem for our subjects in the protocol study (see below).

Overall these findings suggest that provision of an example solution with specific instructions to refer to the example, facilitates some transfer of learning regardless of the addition of a formal definition of the problem types or the separate addition of a mapping relation. However, provision of an example solution, formal definition, and mapping relation, at training, facilitates substantial transfer of learning. The analyses of errors suggested that students when not provided with a formal defintion and mapping relation, focussed their attention to a part of the solution which was largely irrelevant to writing a recursive function. Students provided with this information showed a pattern of errors suggesting that they were attending to the recursive function itself.

CONCLUSIONS

The principle findings of the two studies were as follows: student programmers in solving successive and similar programming problems employ inappropriate mapping strategies from example to current problems. These mappings take place at a surface or syntactic level and involve mapping a solution givens onto a problem statement givens (see also Pirolli & Anderson, 1985). It was not until our students started mapping example problem givens onto current problem statements that we detected any signs of an eventual successful solution. When students are provided with a formal definition of the problem types and a mapping relation which directs their attention to example problem givens and current problem givens, then substantial transfer of learning may be observed.

It seems reasonable to conclude that student programmers are particularly poor at employing examples to aid problem solving. Two factors mitigate against successful analogical problem solving: Firstly, students have little conception of the abstract relations between similar problems (e.g. the recursive function) and, hence, tend to focus on surface similarities. Secondly, and perhaps consequently, students employ inappropriate mapping strategies between the example and current problem.

These impediments to analogical problem solving may be overcome, to some extent, by providing the student with a statement of the abstract relation between problems and a mapping relation which directs the student's attention to commonalities betwen problem givens (rather than solution to solution, or problem to solution). We suggest that this is effective because it facilitates the process of solving proportional analogies. In conclusion, we note that text books and programming manuals rarely provide abstract and mapping relations in their example problems. This is a serious omission. In further research (in preparation), we have found that simply by providing this information in a manual and directing the student to map all parts of the analogy and current problem, subsequent problem solving is greatly facilitated. More importantly, students tutored

in this way show a comparatively fast acquisition of the abstract
relations underlying sequences of programming problems.

REFERENCES

Anderson, J.R., (1983). The architecture of cognition. Harvard
     University Press.

Anderson, J.R., Farrell, R., & Saunders, R., (1984). Learning to
     program in LISP. Cognitive Science, 8, 87-129.

Eisenstadt, M., (1978). Artificial intelligence project. Units 3/4
     of Cognitive Psychology: a third level course. Milton
     Keynes: Open University Press.

Gick, M.L., & Holyoak, K.J., (1980). Analogical Problem Solving,
     Cognitive Psychology, 12, 306-355.

Hayes, J.R. & Simon, H.A., (1974). Understanding written problem
     instructions, in L.W. Gregg (ed) Knowledge and Cognition,
     Lawrence Erlbaum Associates.

Kahney, H., (1982). An in-depth study of the cognitive behaviour of
     novice programmers. Human Cognition Research Laboratory,
     Technical Report No.5.

Kahney, H., (1986). An interpretation theory for analysing verbal
     protocols. Human Cognition Research laboratory, Technical
     Report, in press.

Pirolli, P.L., & Anderson, J.R., (1985). The role of learning from
     examples in the acquisition of recursive programming
     skills. Canadian Journal of Psychology, 39, 240-272.

Reed, S.K., Dempster, A., & Ettinger, M., (1985). Usefulness of
     analogous solutions for solving algebra word problems.
     Journal of Experimental Psychology: Learning, Memory, and
     Cognitiony, 11, N.1, 106-125.

Programming Tools for Prolog Environments

Paul Brna, Alan Bundy, Helen Pain and Liam Lynch

Department of Artificial Intelligence
University of Edinburgh
Scotland

Abstract

The Prolog programmer programs in an environment which provides a number of debugging tools. There is often a mismatch between the way a programmer describes some perceived error and the way in which a debugging tool needs to be used. Worse, there are some problems which existing tools cannot tackle easily —if at all.

The main aim of the work described is to construct a coherent framework on which to base the design of programming tools. This paper describes a particular classification of programming errors. Error classification is then used to provide a natural description of the tools that can, or could, assist the programmer.

Examples are given of useful tools which are not part of well known current Prolog implementations and suggestions are made as to how current tools can be improved to increase their utility.

1 Bugs and Tools

The sources of Prolog programming errors are numerous. Taylor has six levels at which errors may arise (Taylor & duBoulay, 1986). Her analysis is principally concerned with the processes entailed in starting with some problem statement, formalising this problem and eventually producing a Prolog program. Coombs and Stell have investigated the possibility of helping novice programmers detect misconceptions in their understanding of backtracking in Prolog (Coombs & Stell, 1985). Other work, by van Someren, has indicated that several programming errors result in programmers trying to write programs by using familiar concepts from some language other than Prolog (vanSomeren, 1985).

Our current interest, however, lies in the problems that flow from incorrect Prolog programs and the match between the debugging tools provided and the nature of the programming problem.

Our aim is to motivate the construction of new and improved tools based on an analysis of bugs and program debugging. For the moment, the term 'Prolog bug' may be taken to refer to some error that is responsible for the creation of an incorrect Prolog program. This necessarily weak definition is clarified and expanded in what follows.

1.1 Current Prolog Tools

The available tools are divided into *dynamic* and *static* ones. Dynamic tools are applied at run time and are usually bundled together inside a trace package. Static ones are applied at consult time[1].

Recent work at the Open University by Eisenstadt and his workers has focussed on the provision of more useful tools (Eisenstadt et al, 1984; Eisenstadt, 1984). In particular, Eisenstadt has attempted to incorporate both methods of controlling the amount of information

---

[1]Although some tracers provide a post mortem analysis which allows the possibility of easily combining so-called dynamic and static tools

revealed by a dynamic tracer and some knowledge about the kinds of error made by programmers using a post mortem analysis of an extended trace of the computation (Eisenstadt, 1985). Rajan has produced animated tracing tools for novices along with a general set of design principles (Rajan, 1985). The implementation of the transparent Prolog machine —see (Eisenstadt & Brayshaw, 1986)— seeks to combine the ideas found in (Eisenstadt, 1985) and (Rajan, 1985) together with a coherent account of how Prolog works (Bundy et al, 1985).

Other workers have explored Prolog debugging techniques. Shapiro's work is well known (Shapiro, 1983). Lloyd is developing a more principled approach based on extending the Prolog syntax (Lloyd, 1986) while Pereira has extended Shapiro's ideas about *Algorithmic Debugging* to provide a more efficient system (Pereira, 1986). The common idea is to look for missing and wrong solutions in a manner which requires the programmer to guide the search using a dialogue driven by the debugging system.

Lloyd's system has the merit of a clear declarative description of the bugs for which he searches while Pereira's rational debugging system provides a style of debugging that may well be more convenient for the programmer. These extensions to Shapiro's work are important but, as yet, they are based explicitly on a simple classification of errors which needs further development, and a simplified view of debugging strategies[2].

1.2 The Importance of Classifying Bugs

The classification and listing of bugs and debugging strategies is of fundamental importance to give a foundation for motivating new, or improved, tools. In this section, the nature of bug classification is discussed together with the methodology for obtaining it.

1.2.1 Different Levels of Classification

Classification can be at a number of levels. For example, the symptoms which are presented to the user, or the underlying causes for these bugs. Note that a particular symptom may be caused by a chain of causes, so that there is a range of causal explanations from shallow to deep. For instance, the symptom might be that a procedure call seems to be taking a long time. The immediate cause of this might be that the program may be in an infinite loop. This might be because of a programming error —for example, that the body of a clause contains a literal identical to the head. And this might be because the programmer has an underlying misconception about recursion. These four examples are generalised to the four levels: symptom, program misbehaviour, program code error and underlying misconception. These ideas are illustrated below and defined more carefully in section 2.

A classification at the program code error level might include: missing procedure or clause, multiple copies of procedure or clause etc. At the symptom level there is, for example, non termination, error message, etc.

Note also that the same symptom can arise from different underlying causes, or that the same cause can give rise to different symptoms under different circumstances, so that each level will give rise to a different classification. For instance, a procedure call may take a long time because the program is complicated, or looping or inefficient, or because the computer is slow, or has crashed. Misconceptions about recursion can also cause the construction of programs that fail when they they were intended to succeed, or return with the wrong answer, etc.

1.2.2 All Levels Are Potentially Interesting

All these levels are of interest:

- To evaluate dynamic debugging tools it is the symptom and program misbehaviour levels that are of interest. The student who is wondering why his/her procedure call is taking

---

[2]The research has, in the main, been applied to standard Prolog. It is necessary that these approaches be extended to handle new logic programming languages such as PARLOG, CP and GHC.

such a long time must chose a tool mainly on the basis of that symptom, although they might also entertain some hypotheses at the program misbehaviour level.

- To evaluate static debugging tools it is the program code error level that is of interest. For instance, it would be reasonable to expect a tool which diagnosed potential infinite loops —for example, by looking for clauses with heads identical to a body literal. It would be nice if such tools were also capable of giving higher level causes and remedial help to remove misconceptions, but this kind of tool requires fundamental research.

- To advise on teaching methods it is necessary to be interested in the program code error and underlying misconception level. For instance, to find more successful ways to teach recursion so that students did not suffer so readily from misconceptions about it.

- To advise on language improvements/extensions it is necessary to be interested in the program code error and the underlying misconception level. For instance, a language might be suggested in which recursive definitions were only accepted if they were shown to be well-founded (i.e. non-looping).

1.2.3 The Consequences

For any given bug it is likely that there is a hierarchy of strategies corresponding to the different causal levels. This might entail a method of teaching that avoids the misconception, a language improvement that makes the bug impossible, a static tool that detects it at edit or compile time, and a dynamic tool that tracks it down at run time. All these are valuable and should be passed on to teachers, language designers, and tool builders. In the short term the low level, symptom oriented solutions will be most valuable, since they will lead to tools that can patch the current badly engineered language and ill-educated programmers. In the longer term the higher level solutions will be more valuable, since they will lead to languages and teaching methods which avoid problems. However, the world being the imperfect place it is, programmers will always make mistakes and there will always be a role for the debugging tools.

2 Classifications of Prolog Bugs

A multi-layered classification of Prolog bugs is described. This involves four layers. These are: symptom, program misbehaviour, program code error and misconception. The latter level is not considered in any great detail but analyses are given for the other three. The program code error level is further expanded by means of two different classification schemes.

A four-level description is provided. Later, each of the four levels are elaborated separately:

Symptom Description If a programmer believes that something has gone wrong during the execution of some **Prolog** program then there are a limited number of ways of directly describing such evidence. For example:

- exit with Prolog/operating system error
- (apparent) non-termination
- generation of Prolog error message
- unexpected "no" or "yes"
- wrong binding of answer variable
- unexpected generation or failure to generate a side-effect

Program Misbehaviour Description The explanation offered for a symptom. The language used is in terms of the flow of control and relates, therefore, to run-time behaviour. For example, the hypotheses that might be entertained concerning (apparent) non-termination:

- there is a loop
- the computer system is very heavily loaded
- the system has crashed
- the computation takes a very long time

Other potential hypotheses may involve descriptions closely related to the symptom level but at a different level of detail —such as the unwanted success of some subgoal followed by a cut which then causes a subsequent clause not to be used.

Program Code Error Description The explanation offered in terms of the code itself. Such a description may suggest what fix might cure the program misbehaviour. For example:

- there is a missing base case
- there is a clause that should have been deleted
- a test is needed to detect an unusual case

Underlying Misconception Description The fundamental misunderstandings and false beliefs that the programmer may have to overcome in order to come to terms with specific features of the language. For example:

- recursion is really iteration
- if a subgoal fails then the goal fails
- *is/2* is really assignment
- All arithmetic expressions are automatically evaluated

Each of these four levels of error description is now analysed further.

2.1 Symptom Description

It is supposed that the program under investigation is treated as a black box by the programmer for the purposes of describing symptoms. Now 'the program' may well be a subset of a much larger program but that does not matter[3].

In the broader context of programming environments, the use of various tools will be taken into account —such as trace packages and the cross referencer— but it is argued that they cannot make any contribution to the program's *symptom* description as their role is to open up the 'black box' to the programmer's inspection.

At the symptom description level, the classes of event that can be described permit references to:

- Error messages
  - from Prolog
  - from the operating system
  - from an editor which is running Prolog etc.

- Termination issues —this includes:
  - unexpected 'apparent' failure to terminate
  - unexpected termination —which includes the possibility of terminating the Prolog session
  - termination with an unexpected 'value' —for example, a goal succeeds when it should fail

---

[3] This implies that no reference can be made to the program's internal structure.

- Instantiation of Variables which includes:

  - unexpected failure to instantiate a variable
  - unexpected instantiation of a variable
  - a variable is instantiated to an unexpected value

- 'Visible' side effects

  - unexpected failure to produce a specific effect
  - unexpected production of a side effect
  - production of a side effect with an unexpected value

This analysis has imposed an obvious pattern on the possible outcomes with the exception of the entry for error messages. There is no reason, however, why error messages cannot also be included in the same way although it has been pointed out that an 'unexpected failure to produce an error message' may be a rather unusual event. Nevertheless, in the course of determining that some program functions correctly it might be expected that Prolog ought to produce some error message.

Perhaps it is worth pointing out that this analysis does not provide for completely determined error descriptions. In real life, it is expected that there will be a wide range of symptom descriptions[4]. In particular, symptom descriptions may well be clothed in the terminology imported by the programmer from the nature of the programming task being undertaken. This might include terminology from the domain in which the programming task is located, the data structures and algorithms that the programmer has in mind. It is assumed that it is possible to translate such domain specific descriptions into the appropriate Prolog-specific ones. As far as describing the programmer's expectations, a totally separate effort is required to provide a suitable description language.

2.2 Program Misbehaviour

The next stage is to look into the 'black box'.

2.2.1 The Ways in Which a Program May Misbehave

The level of symptom description treats the program as a 'black box' whereas the level of program misbehaviour requires that the black box be opened up.

It is assumed that the programmer knows the names and arities of the various predicates defined. This is consistent with the requirement that each 'black box' has some means of identification or 'handle'.

Whatever is said here must be seen as relative to some Prolog story. A *Prolog Story* is an explanation of the workings of the Prolog interpreter or compiler which a programmer can use to understand and predict the execution of a Prolog program. The basic outline of the 'Proposed Prolog Story' is adopted here (Bundy et al, 1985).

The program can be 'opened up' to examine the program behaviour as it is executed. The program can be seen as a set of connected 'black' boxes. Examining the program behaviour incorporates the symptom description behaviour in section 2.1 but includes an extra aspect which can be termed 'flow of control'.

What is needed is some way of capturing *flow* which suggests that *sequences* are of interest. It is suggested that the principle is adopted that sequences are the transitions from one box to the next. This is the principle that only *local* flow should be considered.

The program can be considered as a set of black boxes:

---

[4]When programmers comment on their code they often mix symptom description with other types of description. This is not the point at issue.

- An expected transition fails to occur

- An unexpected transition occurs

- The expected transition occurs but the instantiation pattern is unexpected

The types of black box are:

- The module —the box consists of an unordered set of predicate definitions. The module can be thought of as a complete program with no undefined predicates. A 'handle' on the box is the principle functor and arity of any procedure visible from 'outside' the module.

- The predicate —the box consists of the (ordered) set of clauses that form the (incomplete) definition of a given predicate. The definition is incomplete if any subgoal depends on any predicate that is not built-in. The handle on the box is the principle functor and its arity.

- A clause —the box is the head and body of a single clause. The handle is the head of the clause.

- An argument of a predicate —the box is the (single) argument while the handle is the name and arity of the argument[5].

The simplest assumption is of the program as a set of clauses. The Byrd box model allows for a program to be a set of predicate definitions but this model has not always been strictly followed by implementors.

2.3   Program Code Error Descriptions

Programming errors can be described at the code level in terms of the syntactic structure of the code itself. For this approach, the principle is adopted that no description must refer to any particular programming techniques being used. For example, consider a faulty version of the standard program for **member/2**:

    member(X,[Y|Z]):-
        member(X,Z).

This could be described by saying that the program code error description is:

    missing base case

but this invokes the means for describing some *programming technique* for a simple recursive procedure.

2.3.1   The First Code Level Classification

The program can be divided up in terms of the kinds of 'black box' discussed at the program misbehaviour level. For division by module, there are three basic cases:

- Missing module: a module that was expected to be present was missing. This could be the consequence of failing to load a module or consult a necessary file.

- Extra module: a module is found which was not expected to be present. This can result from failing to edit out a redundant module.

- Wrong module: some expected module has an error description. This means that the program code error description has to be applied recursively to the module.

---

[5]Provided it is not a variable or a constant in which case there are no further boxes to be opened.

A corresponding set of possibilities exists for division by predicate.

At the clause level, the program is considered as a set of clauses. The possibilities are:

- Missing clause

- Extra clause

- Wrong clause

It is also necessary to capture the sense of order normally required by the standard Prolog search strategy. This means that there is the extra possibility of

- Wrong clause order

At the clause level, the description of a clause can be taken to involve the sub-components of the head of the clause and the body of the clause. The classes of error connected with the head are:

- Missing head

- Extra 'head' (this one is effectively a syntax error)

- Wrong head

The *body* of a clause can be handled in a similar manner but with wrong subgoal order as an additional possibility.

It is now necessary to account for errors at the level of a single term or single subgoal. Given some term, the only way there can be a problem is that the term is the wrong term. The categorisation can be further expanded by sub-dividing terms into atoms, variables and compound terms. This, in turn, has to take into account such errors as: missing predicate name, extra predicate name and wrong name as well as errors associated with the argument list.

The classification of the code level described above can be described as 'syntactic'. That is, it did not reflect any declarative or procedural semantics on the part of the programmer, but merely lists the ways in which an actual Prolog program might differ syntactically from some correct program.

The advantages of this classification scheme are:

- It includes all possible bugs

- It is simple and regular

- There are only a finite number of bug types

The disadvantages are:

- It demands a template giving a detailed and inflexible description of the correct program

- In general, there is no way to know what this template is

2.3.2   A Technique Oriented Classification

Other classifications are both possible and desirable. Another basis for classification is now suggested. It is based on the notion of programming techniques.

An example of such a programming technique is a *failure driven loop*. An analyser for this technique produced by Lynch takes a program and tests it to see whether there is a well formed failure driven loop (Lynch, 1986). The code contains a definition of what one is and looks for violations of this definition.

For instance, a failure driven loop must have a clause that always fails and that contains at least one non-deterministic literal and one that side-effects. Note that this definition includes some procedural semantics: fail, side-effect, etc. Note that it is not as detailed and inflexible as the template of the syntactic classification. There could be other clauses and literals. The non-deterministic and side-effecting one could be the same or different.

The various ways in which a program can violate the definition of a failure driven loop constitute a technique-oriented bug classification, namely:

- No failing clause

- No side-effecting literal

- No non-deterministic literal

The differences between this technique-oriented classification and the syntactic one are:

- This one does not cover all possible bugs.

- There are an indefinite number of techniques, and hence an indefinite number of bug types.

- This classification might get arbitrarily complex.

- This one requires a flexible definition of a technique rather than a detailed and inflexible template.

- One might more readily know what definition was intended than what template was intended.

Various program analysers are being built by Liam Lynch based on both the above classifications (Lynch, 1986). His 'tail recursion' analyser uses a fixed template and produces a bug analysis drawn from our syntactic classification. He is now extending this to a syntactic analyser which can take any template and produce an analysis based on the full syntactic classification.

His 'failure driven loop' analyser uses a definition of this technique and produces a technique-oriented bug analysis. He is building similar analysers for other techniques. For instance, a technique-oriented tail recursion analyser, which would contrast with his existing syntactic one and illustrate the difference between the two classification schemata.

The syntactic analyser will have the virtues of generality but the limitations of inflexibility. In the context of an automated programming tutor, such an analyser will be able to criticise any student program, provided the teacher has provided a template but will be liable to reject perfectly good answers just because they use a solution method slightly different from that intended by the teacher.

The technique analysers will have the opposite properties. They will only be able to criticise programs that are intended to be examples of some technique, but will be able to accept a wide range of correct solutions —for example, any correct failure driven loop. However, they will only be able to criticise the program on the basis of its fit to the technique, not on other grounds —such as producing the wrong answer!

Such technique analysers are potentially useful debugging aids in a Prolog environment. The programmer could ask that a program be tested for 'tail-recursiveness' etc. Such analysers fit into the range of other static tools like mode-finders, type-checkers, etc. Note that the syntactic analyser could not be used in this way because there is no way that users could want their programs checked if they already knew the template they were to fit.

Other techniques are being considered in order to build up other technique-oriented bug classifications. In particular, the techniques of building up recursive data structures by pattern matching in the clause head, and building them up in the clause body with an accumulator.

Other related work at Edinburgh takes a slightly different line. Chee-Kit Looi is seeking to harness a number of methods — such as symbolic evaluation, mode checking, type checking etc. — to build a Prolog Intelligent Tutoring System (Looi, 1986; Looi & Ross, 1986).

2.4 Misconception Description

As yet, no attempt has been made to produce some higher level bug classifications in terms of misconceptions. To uncover the deeper bugs we will need to take into account on-going research into the misconceptions of novice Prolog programmers such as that by Taylor and van Someren (Taylor & duBoulay, 1986; vanSomeren, 1985). It is expected that some empirical studies will be necessary to uncover misconceptions and link them with the lower level bugs already identified.

3 Debugging Strategies in Prolog

In previous sections the classification of Prolog bugs was considered at each of four levels: symptom, program misbehaviour, program code error and misconception: Here the beginnings of an investigation into debugging strategies are described in terms of the first two levels of description.

A Prolog programmer's first indication of a run time bug is one of the types of symptom described in section 2. Typically, s/he will then try to identify a bug at the program misbehaviour level to account for this symptom, and hence to a code error which can be corrected[6].

The focus is now the step from a symptom level description to a program misbehaviour level description. For each error type, the question is asked: "Using existing environmental tools, do efficient bug avoiding or debugging strategies exist?" If the answer is no, the next question is "can better dynamic or static tools be suggested?"

Since the immediate symptoms of run time bugs are under investigation then most debugging strategies will involve the tracer and the discussion will be almost exclusively procedural. The role of declarative solutions is in static bug avoiding tools or programming and teaching strategies.

The bugs associated with 'exit with Prolog/operating system error' will not be discussed here.

3.1 Current Tools for Inspecting Program Behaviour

In some sense there is only one possible tool available for looking at run-time program behaviour —the trace package. Therefore the trace package can be switched on, the execution of some n port model[7] followed or a 'spy' point set etc. A number of other tools such as XREF (cross referencer) and MODGEN (mode generator) are potentially useful —and there has been some consideration as to whether they can be tied into the trace package— but they generally apply to the (static) code rather than the (dynamic) run-time behaviour.

The 'standard' trace package is not a uniform entity. A 'trace package' is a collection of facilities built to form a coherent whole. The tracer provides facilities to enable the programmer to 'walk over' some abstraction of the program's behaviour[8] —often a tree-like representation— and provide various possibilities at each node in the representation. Through an analysis of the ways in which a program can misbehave it is possible to identify (at least some of) the gaps in the provisions for each environment.

3.2 (Apparent) Non-Termination

Suppose the program seems to be taking a long time to return. The problem might be:

---

[6]However, at AAAI-86, an example was given of going straight from symptom to code error. Given an unexpected "no" the programmer would immediately turn on an 'undefined procedure' mode and look for one.

[7]Where n=4 for Quintus, DEC-10 Prolog, n=3 for micro-Prolog and n=7 for Dave Plummer's SODA debugger —(Plummer, 1986).

[8]Such as the execution AND/OR tree, the AND/OR 'search space', Eisenstadt's AORTA diagrams (Eisenstadt & Brayshaw, 1986) or the Byrd Box execution model.

1. A loop

2. The computer is slow today

3. The computer has died

4. The program just takes a long time

Items two and three can probably be tested by some operating system command. Consideration is now given to testing item one.

In most Prologs the only way to test for non-termination is to interrupt the program run and then use the tracer to look for repeated goals, etc. This is most inefficient. The combination of inefficiency and uncertainty makes this one of the the the worst bugs to diagnose.

One exception is Logicware's MProlog[9]. It has a user defined exception handling facility which allows one to recover gracefully from most bugs. This has a default setting to go into a break state at depth 10k. Often, such a symptom suggests a loop. To track it down the programmer can enter the tracer and search around for duplicated goals, etc. The MProlog tracer allows tracing back from this break point. This is useful to find out why Prolog got into the current state.

A useful bug avoiding tool would be a static analyser to detect potential loops in code. Several groups either have or are working on such tools[10]. Such static tools are useful but a dynamic tool can handle situations that the static tool can never detect.

A dynamic tool to search for loops at run time would therefore be most useful. Shapiro, for example, makes use of a depth-bound Prolog interpreter to search for simple loops in PDS (Shapiro, 1983).

More generally, a subsumption checker would go a long way to meeting this need. Subsumption checking is very expensive in general, so one would want to be able to turn it on and off at will, i.e. Prolog should be runnable in subsumption mode. It might be useful to have it on automatically after some prearranged time or at some prearranged point in the program in order not to waste time on checking bits of the search space suspected to be correct. Such a tool would be more useful than the MProlog depth bound, because it would locate the problem more accurately and involve the programmer in less search.

Checking for non-termination is an undecidable problem in general, so no perfect tool could be provided — neither a static nor a dynamic one. Improvements would always be possible and the programmer would always have to be prepared to resort to 'hand' checking of the tracer output. However, most non-terminations are caused by simple goal repetition, so a simple loop checker would go a long way.

3.3   Prolog Error Messages

By its very nature, Prolog provides few error messages. Typically, error messages are mostly for inappropriate arguments to system predicates, e.g. =.., is, etc, or for exhaustion of some system resource, e.g. stack overflow. Some errors are fatal and some not.

For system resource exhaustion errors one is mainly concerned to distinguish program looping from 'genuine' exhaustion. Looping tests were discussed in the last section and will not be repeated here. A major problem with this kind of error is that they are likely to make it difficult to continue to run Prolog and thus to provide any Prolog debugging tools. Something like the MProlog solution seems to be required, i.e. of predicting a problem before the resource is actually exhausted and going into a break to allow investigation.

In many Prologs the standard way to test for system predicate bugs is to discover which system predicate they refer to, and which procedures call this predicate, and then to trace these

---

[9]Only an informal investigation has been made of the facilities provided by a number of Prolog implementations. Some of the following is based on discussions with Prolog suppliers at AAAI-86. Further work has yet to be done to detail both strengths and weaknesses of various systems.

[10]For example, McCabe and his colleagues incorporate the idea in their Alvey proposal (McCabe et al, 1986).

procedures to try to spot which one is at fault. If many procedures call the offending system predicate, or if any of these procedures is itself called often, then this is a most inefficient debugging strategy.

Since the Prolog interpreter 'knows' exactly which procedure call caused the error message to be printed, it should be simple to arrange for the tracer to be invoked at precisely the right point. In fact, MProlog allows just this but this is not the case with either Quintus Prolog v1.6 or Edinburgh Prolog (NIP).

Having found the offending system predicate call, the programmer will then probably want to trace back up the program run, looking for the point at which the trouble started. Most tracers[11] do not currently allow this: 'redoing' only being possible back to where the tracer was invoked. The 'debug' mode in Edinburgh Prolog or Quintus ensures that information is kept which can be used to provide such facilities.

This search backwards for the point of trouble could also be automated, e.g. an uninstantiated input to 'is' might be identified and the tracer requested to backtrack to where this was first introduced. Back-searches for the point of variable introduction or binding is a general requirement.

3.4   Side-Effects

Unexpected generation of side-effects is much like system predicate error messages — it is desirable to get into the tracer at the point at which the side effect happened and then root around. The main difference is that Prolog cannot be expected to do this without the programmer specifying which side-effects are regarded as unexpected and unwanted. Some way is needed of stating that writing this term or asserting that one is unwanted and then to be put into the tracer at the point at which these events are happening.

Failure to produce side effects can be tackled initially by spying the procedure which should have produced the side effect. If this fails because this procedure was never called then it is a case such as 'unexpected no' and the remarks in the next section can be adapted.

3.5   Unexpected 'No' or 'Yes'

Prolog has been called "the language that likes to say no". 'No' when it was expecting 'yes' or some output binding is certainly one of the most common symptoms of trouble. It is also one of the hardest to debug, since there are so many potential causes and so little evidence to go on.

The normal debugging technique is to use the tracer to compare the actual search space with the one expected. The main goal has failed and one is trying to trace that failure down through the search space to the lowest failing subgoal. This may be caused by:

- The unexpected failure of some clause

- The unwanted success of some clause — with a cut then causing a subsequent clause not to be called

- A missing clause

The programmer might suspect some particular procedure to be at fault, e.g. if it has just been added to or edited in a previously correct program. In this case s/he will spy this procedure. Failing such suspicions, the programmer must resort to a more exhaustive search. Most programmers either step through the program top down, or test each subprocedure in turn working bottom up, or try some middle out variant.

Exactly where the tracer should initially enter in the execution history is a difficult problem. Shapiro offers a solution which depends crucially on keeping track of the programmer's

---

[11]MProlog, again, being a rare exception.

expectations (Shapiro, 1983). We assume here that these are unavailable at the point at which the tracer is invoked.

One of the simplest possibilities is to enter the tracer at the point at which the first goal failed in the chain of most recently failing goals. Either this goal should not have failed and selective backwards tracing can now be attempted to pinpoint why, or this goal should never have been called and selective backwards tracing can now be used to discover how this part of the search space was reached. Note that this technique requires the ability to re-enter the tracer after a top level call has been exited — no Prolog system currently provides such an ability[12] but it should not be hard to implement provided one is prepared to go into a special mode before making the call.

Unexpected 'yes' when 'no' was expected is very rare but can be dealt with in a similar way to the above.

3.6   Wrong Binding

The wrong binding of an answer variable is currently similar to the previous case of unexpected 'no'. However, the additional information implicit in the erroneous binding gives us much more to go on. The tracer could be made to back up in turn through each unification which contributed to the binding. At each stage it is known what the binding was before the procedure call and what new instantiations were made[13]. The user can then choose whether to go on up or root around at this point. Note that this requires the ability (originally suggested in the 'error messages' section) to trace variable bindings, as well as the ability (originally suggested in the 'unexpected no' section) to trace back into a terminated program.

It has been suggested that answer variables could be traced forwards instead of backwards. This avoids the problem of tracing backwards and re-entering terminated programs, but it is less efficient as a debugging strategy. Forwards tracing forces the programmer to enter and search bits of the search tree which will ultimately fail, whereas backwards tracing will enter only the ultimately successful branch of the tree.

This completes the survey of the symptom level bugs descriptions. Now, attention is given to the second level: program misbehaviour.

3.7   Program Misbehaviour

Debugging at the program misbehaviour level consists mainly of tracking the immediate cause of the symptom to its 'initial' cause[14]. When the 'initial' cause is found it can be identified because its cause can be found in some corresponding code level bug. For instance, the immediate cause may be some Prolog error message. This might be caused by an unbound variable, caused by a clause being unexpectedly called, caused by an earlier clause unexpectedly failing, caused by its head not matching the goal. This last is the 'initial' cause at the program misbehaviour level because its cause is a code level bug, say a mistyped function name.

It is assumed that the test for an 'initial' cause can be done by inspection of the code corresponding to the erring clause, i.e. without any special tools beyond the ability to access source code from the tracer [15] However, the tracer might be augmented to assist in the tracking of the causal chain between immediate and 'initial'.

Most such augmentations have already been mentioned above:

- The ability to search for nested identical goals or, more generally, subsumption

- The ability to trace backwards ('redo') from a break point

---

[12] To the best of our knowledge.

[13] Pereira uses this kind of term dependency information in his oracle-based debugging system (Pereira, 1986).

[14] The term 'initial' is odd because, as will be seen, it is initial only at the program misbehaviour level and itself has a cause at the code level. That is why it is in, and will remain in, scare quotes.

[15] Sadly lacking in most tracers. A rare exception is Dave Plummer's SODA debugger (Plummer, 1986).

- The ability to trace back into a terminated program

- The ability to name a variable and have the tracer look for its last binding or its introduction

- The ability to drop into the tracer at the point where an error message or unwanted side effect were generated

The reason that most of these augmentations have already been considered is that, for that most part, the program misbehaviour bug classification echoes the symptom level one but within the search space rather than external to it. The main exception is unexpected calling or non-calling of a clause, but this does not seem to require any augmentations or tools beyond those discussed above.

4 Conclusion

Examination of available debugging strategies for each of the bug types at the symptom level has been a fruitful activity in terms of revealing the shortcomings of existing Prolog environments and suggesting improvements. The normal debugging strategies available in DEC-10/Quintus type environments are very inefficient for many bug types, although the situation is a little better in some other Prologs, e.g. ESI's Prolog-2 and especially Logicware's MProlog.

A little thought suggests some dramatic improvements over the current situation: in the main improvements to the tracer to make it more selective and to enable kinds of tracing not currently allowed. These suggestions are listed at the end of the previous section. These could be implemented, although some of them would be costly in resources and one would want to switch them on specially rather than always pay the overhead. The proposal by McCabe, Wilk, Thwaites and Ramsay indicates an intention to provide tools to do a post mortem analysis of the stack which would make implementation of some of the above ideas feasible (McCabe et al, 1986).

The debugging strategies used by Prolog experts for each Prolog bug symptom and for bugs at other levels will be investigated further. More ideas will emerge as further Prolog bug types are considered [16] and further relations between them. More work is especially needed on the code level and static tools, and about declarative classifications and tools.

Much work is still needed to develop the classification of program code errors. In the near future, there will be a more thorough investigation of the range of programming techniques.

Acknowledgements

This research was supported by Alvey Grant number GR/D/44287. We thank the other members of the Mathematical Reasoning Group and the Programming Support Group at Edinburgh for numerous conversations and useful feedback. Our thanks for conversations at the AAAI-86 exhibition with: Jonathan Grayson and Alex Goodall (Expert Systems International), Lew Baxter (Logicware), and Fred Malouf and Bill Kornfeld (Quintus).

References

Bundy, A., Pain, H., Brna, P. and Lynch, L. (1985). *A Proposed Prolog Story*. Research Paper 283, Dept. of Artificial Intelligence, Edinburgh.

Coombs, M. J. and Stell, J. G. (1985). *A Model for Debugging PROLOG by Symbolic Execution: The Separation of Specification and Procedure*. Research Report MMIGR137, Department of Computer Science, University of Strathclyde.

---

[16]It is suspected that it may be necessary to refine even the symptom level further.

Eisenstadt, M. and Brayshaw, M. (1986). *The Transparent Prolog Machine TPM*. Technical Report 21, Human Cognition Research Laboratory, The Open University.

Eisenstadt, M. (1984). A powerful Prolog trace package. In O'Shea, T., (ed.), *ECAI-84: Advances in Artificial Intelligence*, Elsevier Science Publishers.

Eisenstadt, M. (1985). Retrospective zooming: a knowledge based tracking and debugging methodology for logic programming. In Joshi, A., (ed.), *Proceedings of the 9th International Joint Conference on Artificial Intelligence*.

Eisenstadt, M., Hasemer, T. and Kriwaczek, F. (1984). An improved user interface for PROLOG. INTERACT-84, IFIP conference on Human-Computer Interaction.

Lloyd, J.W. (1986). *Declarative Error Diagnosis*. Technical Report 86/3, Department of Computer Science, University of Melbourne.

Looi, C.K. and Ross, P.M. (1986). Automatic program debugging for a Prolog Intelligent Tutoring System. Submitted for Publication.

Looi, C.K. (1986). *Automatic Program Debugging for a Prolog Intelligent Teaching System —A Thesis Proposal*. Discussion Paper 30, Dept. of Artificial Intelligence, Edinburgh.

Lynch, L. (1986). *A Thesis Proposal for a Computer Progam to Teach Prolog*. Discussion Paper 10, Dept. of Artificial Intelligence, Edinburgh.

McCabe, F., Thwaites, G., Ramsay, A. and Wilk, P.F. (1986). Logic Programming Environment. A Submission to the Alvey IKBS Directorate.

Pereira, L.M. (1986). Rational debugging in logic programming. In Shapiro, E., (ed.), *Third International Conference on Logic Programming*, Springer Verlag, Lecture Notes in Computer Science No. 225.

Plummer, D. (1986). *SODA: Screen Oriented Debugging Aid*. Software Report 3, Dept. of Artificial Intelligence, Edinburgh, Previously referenced as blue book note 260.

Rajan, T. (1985). *APT: The Design of Animated Tracing Tools for Novice Programmers*. Technical Report 15, HCRL, Open University.

Shapiro, E. Y. (1983). *Algorithmic Program Debugging*. MIT Press.

Taylor, J. and du Boulay, J.B.H. (1986). Why novices may find learning Prolog hard. In Rutkowska, J. and Crook, C., (eds.), *The Child and the Computer: Issues for Developmental Psychology*, Wiley, forthcoming.

van Someren, M. W. (1985). *Beginners Problems in Learning Prolog*. Memorandum 54, Department of Experimental Psychology, University of Amsterdam.

Reasoning About Belief

The Subjective Ascription of Belief to Agents

Afzal Ballim

Box 3CRL, Computing Research Laboratory,
New Mexico State University,
Las Cruces, NM 88003, U.S.A.

CSNET: afzal@nmsu

ABSTRACT

A computational model for determining an agent's beliefs from the viewpoint of an agent known as the system is described. The model is based on the earlier work of (Wilks & Bien, 1979; 1983) which argues for a method of dynamically constructing nested points of view from the beliefs that the system holds. This paper extends their work by examining problems involved in ascribing beliefs called meta-beliefs to agents, and by developing a representation to handle these problems. The representation is used in *ViewGen*, a computer program which generates viewpoints.

1. Introduction

This paper describes a computational model for determining agents' beliefs from the viewpoint of another agent. The model is based on the earlier work of (Wilks & Bien, 1979; 1983), which argues for a method of dynamically constructing nested viewpoints from the beliefs that the system holds; for example, generating what the system believes Frank believes Tom believes about terrorism. Nested belief structures are important in natural language comprehension, planning, and other areas in Artificial Intelligence (Wilks & Bien, 1979; 1983).

We are concerned, in this paper, with the problems involved in ascribing the beliefs of one agent to another agent. The details of generating complicated nestings are discussed elsewhere (Ballim, 1986; Wilks & Ballim, in press). In particular we are concerned with representing, and ascribing a special class of beliefs known as atypical beliefs.

Unlike many other belief systems, this system is subjective, meaning the system does not have veridical access to other agents' beliefs. Instead the system has a set of beliefs, its own, and must infer the beliefs of other agents from this set. Hence there is no appeal to an omniscient observer to aid in resolving problems.

The layout of the paper is as follows: Section 2 presents the background to this work, including a brief description of *ViewGen*, a computer program that generates nested viewpoints; section 3 describes some of the problems involved in ascribing belief to an agent; section 4 presents a taxonomy of meta-beliefs and a representational scheme to handle the problems discussed in section 3; finally, section 5 summarises the paper and indicates areas of further development.

2. Background

The importance of nested beliefs to processes of understanding conversation has only recently been realised within Artificial Intelligence notably in the research of the Toronto group (Cohen, 1978; Allen & Perrault, 1978). Their work, however, has several drawbacks in that points of view are pre-computed, and that the modelling of individuals mentioned in the conversation is neglected. Also, they are more concerned with speech acts and planning. While the relation of our work to speech acts and planning is recognised (an intended use of *ViewGen* is in cooperative planning in a multi-actor system) it is felt that the claims of this paper may be made independently of both.

(Wilks & Bien, 1979; 1983) propose a dynamic model of nested viewpoint generation; this model and its implementation (*ViewGen*) are described in (Ballim, 1986), and in (Wilks & Ballim, in press). *ViewGen* is a computer program which represents and manipulates the beliefs of an agent known as "the system". Given the system's beliefs, it is capable of generating nested viewpoints of what the system believes other agents believe. *ViewGen* only holds minimal representations for what the system believes are the beliefs of other agents. A complex viewpoint is generated by attempting to ascribe the system's beliefs to the agent(s) in question. This is done by taking a relevant subset of the system's beliefs and assuming the agent to have the same beliefs *unless there is evidence to the contrary.*

We write beliefs here as propositions in a language similar to first order predicate calculus (henceforth FOPC) although *ViewGen* represents beliefs in a meta-language known as **FOLSE** (First Order Logic with Sets and Environments). So we may write the belief that the Earth is round as:

$$shaped(Earth,round)$$

or as:

$$Earth\ shaped\ round$$

depending on convenience[1].

FOLSE is similar to logics developed by (Moore, 1980), (Weyhrauch, 1980), (Konolige, 1984), (Levesque, 1984), (Haas, 1986), and (Maida, 1986). The belief that the Earth is round is written in FOLSE as:

wff([shaped1,earth1, round2])

where the numbers are used to represent specific senses of the terms which they follow. Environments are used to represent viewpoints in FOLSE. An environment is a tuple $E = (A, \Psi)$ where A is a sequence of agents and Ψ is a set of propositions[2]. If $A = \langle i_0, i_1, \cdots, i_n \rangle$ and $p \epsilon \Psi$ then this is equivalent to $B_{i_0}(B_{i_1}(...(B_{i_n}p)...))$, where $B_{i_k}p$ can be interpreted as agent i_k believes p. This may be phrased as i_0 believes that i_1 believes that ... i_n believes p.

An example will illustrate the processes involved in *ViewGen*. Suppose the system monitors a conversation between two agents, known as Sally and Frank, concerning the like, or dislike, that they have for each other. A set of beliefs, which is relevant to the discussion, is extracted from the system's beliefs, and an environment is formed from them. This environment contains the beliefs that Sally likes Frank, and that Frank believes Sally dislikes him.

---

[1]   These propositions are well-formed formulas (henceforth wff). A wff becomes a belief when it is asserted as being believed by an agent. Hence the only wffs which are believed, by an agent, are the beliefs that are asserted as being believed by the agent. Furthermore, an agent is only deemed to be aware of those wffs which he believes.

[2]   This is a simplification. Elements of Ψ may be either wffs, or else environments. In the case where it is an environment we may out factor this environment. For example, if $E_0 = (A_0, \Psi_0)$, $E_1 \epsilon \Psi_0$, and $E_1 = (A_1, \Psi_1)$ then we may form $E_2 = (A_2, \Psi_1)$ such that A_2 is A_0 concatenated with A_1 ($A_2 = A_0 \bullet A_1$). Concatenation joins two sequences, hence if $A_0 = \langle i_0, i_1 \rangle$ and $A_1 = \langle i_2, i_3 \rangle$, then $A_0 \bullet A_1 = \langle i_0, i_1, i_2, i_3 \rangle$.

```
?- show_belief([system], [ ], _).

  bel([system],
      [
         wff([likes1, sally1, frank1]),
         bel([frank1],
             [
                wff([dislikes1, sally1, frank1])
             ])
      ])

Yes
```

Figure 1. The Relevant Subset of Beliefs

In the environment (see figure 1) A is the sequence "[system]", and Ψ is a set consisting of two objects; the belief that Sally likes Frank, and another environment which contains the belief that the system believes that Frank believes that Sally dislikes him.

What the system believes Sally believes is not represented here because the system believes Sally to hold the same belief as itself (that Sally likes Frank). The default rule will allow construction of the nesting, of what the system believes Sally believes, such that the system's belief is ascribed to Sally (figure 2).

```
?- viewgen([system,sally1],[],_).

  bel([system,sally1],
      [
         wff([likes1, sally1, frank1]),
         bel([frank1],
             [
                wff([dislikes1, sally1, frank1])
             ])
      ])

Yes
```

Figure 2. The System's View of Sally's View of the Relevant Subset of Beliefs.

Here *ViewGen* has constructed the environment of what the system believes Sally believes. Both the belief, that Sally likes Frank, and that Frank believes Sally dislikes him, have been ascribed to Sally. Figure 3 shows what the system believes Frank believes.

```
| ?- viewgen([system, frank1], [], _).
|
|    bel([system, frank1],
|         [
|             wff([dislikes1, sally1, frank1])
|         ])
|
| | Yes
```

Figure 3. The System's View of Frank's View of the Relevant Subset of Beliefs.

We can see in figure 3 that the system believes that Frank believes Sally dislikes him[3]. This example demonstrates a very simple case. Most cases are more complex than this and present various problems.

3. Ascribing Belief

Thus far we have shown how the default rule can ascribe belief to an agent if there is no further information (as in ascribing the belief that Sally dislikes Frank, to Sally) and how the rule can be overridden by an *a priori* belief (as in the belief that Sally dislikes Frank overriding the belief that Sally likes Frank in the example in figure 3). We now consider more complex cases.

3.1. Blocking Beliefs that an Agent is Unaware of

Ascribing belief should be blocked in cases where a belief exists that an agent does not know something, or does not have a belief on some topic. Consider the scope of negation on the *"know"* operator. Given the following:

$$Smith \quad NOT\text{-}know \quad p \tag{1}$$

$$Smith \quad know \quad NOT\text{-}p \tag{2}$$

It is a parsing/translation issue to decide which form an input should take but representationally (1) implies lack of knowledge while (2) implies knowledge of a negative proposition, hence (1) should **not** be ascribed to Smith, while (2) should be (cf. Wilks, 1986).

We feel that this also holds true for *beliefs*. So for the following:

$$Smith \quad NOT\text{-}believe \quad p \tag{3}$$

$$Smith \quad believe \quad NOT\text{-}p \tag{4}$$

(3) should not be ascribed while (4) should be. In other words, NOT-know and NOT-believe are explicit counter-evidence that we should not ascribe a certain belief to a

---

[3] The system has a set of meaning postulates for terms, and predicates used in propositions. These meaning postulates allow the system to determine that Sally liking Frank is opposite in meaning to Sally disliking Frank. They also contain a partial ordering of agents into classes. One use of this ordering is explained in section 4.

certain agent.

Furthermore, example (1) is more complex than it appears. If the system believes that Smith does not know proposition p then the implication is that the system believes[4] that it does know p. Not only must (1) not be promoted, but it must also cancel the promotion of the system belief that it knows p.

3.2. Atypical Beliefs

We now turn our attention to a class of beliefs, called atypical beliefs, which need a rule which is the opposite of the default rule for ascribing belief.

An atypical belief is a belief which is held by an agent but would not generally be held by other agents. The class of atypical beliefs covers such areas as self knowledge, secrets, expertise and knowledge of uncommon domains (such as the believer's hobbies, skills, personal medical history, etc.).

So, for example, the belief that the Earth is flat is considered atypical. An important point must be made here. In terms of a specific agent the foregoing definition is insufficient. For a belief to be considered atypical, with respect to an agent, the agent must believe the belief to be atypical, i.e., I may believe you to have an atypical belief, however, I may also believe that you think it is a typically held belief. So while I believe it to be atypical, you believe it to be typical.

Expert belief because uncommonly held is a type of atypical belief that presents special problems. Consider the case of medical knowledge. It is possible to reason about a treatment for an illness without knowing the details of the treatment. We may discuss a cure for tuberculosis without knowing the cure. This example shows that simply preventing beliefs on expertise from being ascribed to an agent is often undesirable, even though the agent is not believed to be expert on the topic of the beliefs.

Another type of atypical belief is belief which covers uncommon domains. An uncommon domain is a topic about which a particular agent has a very detailed set of beliefs that most other agents are not aware of; for example, beliefs about a particular agent's childhood. Beliefs on uncommon domains are similar to beliefs on expert domains because both types are uncommon. A belief about some agent's phone number can be treated as a belief on expertise. The experts are those people who know what the agent's phone number is, but non-experts are still capable of reasoning about the agent's phone number.

For the class of atypical beliefs the rule should be **not** to ascribe unless one has explicit evidence to justify ascribing the belief. The problem is one of representing and handling a wide range of types of atypical beliefs. The introduction of meta-beliefs is one possible solution, because meta-beliefs enable explicit representation of atypical beliefs (i.e., for atypical belief p, have the meta-belief $atypical(p)$). Due to the wide range of atypical beliefs, and to the problems that they pose, we use a special representation to handle them. Section 4 discusses this representation.

4. A Representational System for Atypical Belief

McCarthy and others have suggested that lambda expressions be used to represent knowledge of values. (Wilks, 1986) has proposed that expertise may be expressed within a system by use of lambda expressions with restrictions on the capable evaluators of each such lambda expression. Knowing or having the belief represented by the lambda expression means that the agent is capable of evaluating the expression. So the representation for a cure for tuberculosis would be:

---

[4] This implication is valid where p is a proposition of the form **knowing that**, but would not necessarily be valid for p if p is of a **knowing value** or a **knowing skill** form.

$$(CURE\text{-}FOR \quad Tuberculosis) \quad BE \quad (\lambda(x) \ (CURE\text{-}FOR\text{-}TB \ x)) <MDs>$$

The only capable evaluators are those known to be MDs (medical doctors). *ViewGen* uses an extended form of these lambda expressions to represent atypical belief.

4.1. A Taxonomy of Meta-Belief

The form of representation suggested by Wilks allows expressions which can only be evaluated by specific agents or classes of agents. This is a first step towards representing atypical belief, however, it is insufficient.

Consider the major factor that makes a belief atypical; some agent believes that a belief held by another agent is not commonly held. A belief about another belief is known as a meta-belief. It is the meta-beliefs about a belief that mark a belief as atypical. An atypical belief may have a large number of meta-beliefs associated with it.

These meta-beliefs can often be classified according to the relation that they define between an agent and the belief with which they are concerned. We propose a taxonomy of meta-belief (figure 4) and claim that a system which can represent a belief and the associated meta-beliefs (of the taxonomy), is very powerful.

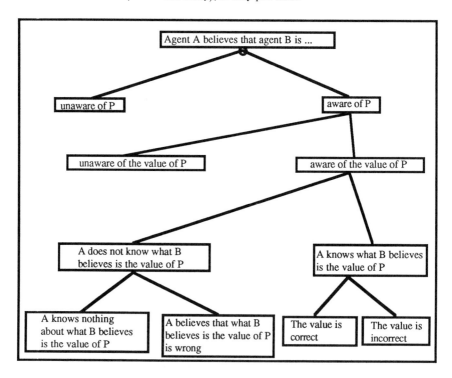

Figure 4. A Taxonomy of Meta-Belief

Consider a belief about an agent's phone number (call this belief P), i.e., the belief is that the agent's phone number is some value. An agent (*A*) may have a number of meta-beliefs concerning this belief P. In figure 4 we see that agent *A* may believe another agent (*B*)[5] to be either aware or else unaware of the belief (P). Somebody who does not know John can hardly have beliefs about John's phone number, and so that person should be unaware of any beliefs about John's phone number.

Further types of meta-belief are possible if A believes B is aware of P. A may believe that B is unaware of the value of P. For example, I may believe that you know that John has a phone number but that you do not know what his phone number is.

If agent A believes that agent B does know the value of P, then A may either know what B believes is the value, or else A may not know what B believes is the value of P. I may know that you believe John's phone number to be "505-526-5444", or I may simply not know what you believe is his phone number.

If A knows what B believes is the value of P, then A may decide whether B is correct or incorrect in his belief, according to what A believes is the correct value, i.e., if I believe that John's phone number is 505-525-5444, and I believe that you believe it's 505-526-5444, then I believe that you are incorrect in your belief about John's phone number.

If A does not know what B believes is the value of P, it is possible that A can use other information to determine whether B is correct or incorrect in his belief about the value of P. For example, if you have not seen John in years, and he has moved house since you last saw him (but you do not know that he has moved house) although you believe that you know his phone number, I believe that you have his phone number wrong. I can believe this even if I do not know what his previous phone number was.

The ability to represent meta-beliefs, of the type described above, is crucial to the process of ascribing belief. *Lambda Formulas* are an extension of lambda expressions which allow the meta-beliefs of a belief to be represented with their associated belief.

4.2. Lambda Formulas

Lambda expressions are finite partial functions. They are computable and map into a finite range of values. We extend the lambda expression representation to a Lambda Formula so we can represent opposing views on the value of such an expression[6]. For example,

$$(PHONE-NO-OF\ John)\quad Be\quad (\lambda(x)(PHONE-NO-OF\ x)\ John) < John >$$

represents an expression which evaluates to John's phone number for John, but is unevaluable to any other agent. The imposition of John as a capable evaluator of the lambda expression (henceforth λ-expression) is equivalent to adding an extra parameter to the expression; the extra parameter being the agent performing the evaluation.

The list of capable evaluators is not necessarily a constant because our beliefs may vary regarding who are the capable evaluators of an expression. Also different evaluators may evaluate the expression to different values, which is also subject to each individual's beliefs.

We want to express different agents as being capable of evaluating such an expression but to different values. We propose a more complex form of the capable evaluators list proposed by (Wilks, 1986). A typical expression in the augmented representation is shown in example 5.

---

[5] The other agent may be A himself. In that case the meta-belief is one of introspection, however it would be absurd for agent A to have a belief that he is unaware of belief P so this situation is not allowed. Introspection is discussed in detail in (Konolige, 1985) and in (Maida, 1986).

[6] It may appear that talking about the value of a proposition is only valid for knowledge of values, such as phone number, etc., but not for propositions of the form "*John is here*"; however, these propositions may be treated in one of two ways: (a) they may often be rewritten to reflect a value (e.g., location of John is x), or (b) the truth or falsity of the proposition may be treated as its value. Coupling the latter case with *unknown* values (see section 4.3) provides us with the power of a Kleene three-valued logic.

$$(CURE\text{-}FOR\ Tuberculosis)\ \ BE\ \ (\lambda(x)\ (CURE\text{-}FOR\text{-}TB\ x))$$

$$\begin{bmatrix} < \{MDs,\ John\},\ v_0 >, \\ < \{Dan,\ Tibetan\text{-}Priest\},\ v_1 >, \\ . \\ < \{Sally,\ Paul\},\ v_n >, \end{bmatrix} \tag{5}$$

The above example contains n sets of capable evaluators and the values to which they evaluate the expression. The basic set of capable evaluators has been replaced by a set of pairs, consisting of a set of capable evaluators and the value that they return. A typical entry, for example, $< \{Sally,\ Paul\},\ v_n >$, is called a Capable_Evaluators-Value pair (abbreviated to CEV pair). It is stipulated that each value v_i must be of the same structure and that the λ-expression returns this structure.

The set of CEV pairs is a set of meta-beliefs about the belief represented by the λ-expression. The CEV pairs in example 5 represent the meta-beliefs that the system believes the agents in those CEVs to be aware of there being a cure for tuberculosis, that they know a cure for tuberculosis, and that the system knows what they each believe is the cure for tuberculosis. Only one of the values $v_0,\ \ldots,\ v_n$ is believed by the system to be the correct value. The other values are considered incorrect. This covers the section "*A knows what B believes is the value of P*" in figure 4.

The set of CEV pairs is also a function table representation of the λ-expression. We refer to the λ-expression and its function table as a λ-formula.

Formally, a λ-formula is a tuple $\Lambda = (\ \lambda,\ \Gamma)$, where λ is a λ-expression and Γ is a function table. A function table Γ is defined to be a set of tuples of the form $\Xi = (\ \Delta,\ \upsilon)$, where Δ is a set of agents, known as capable evaluators, and υ is a value returned by the λ-expression.

4.3. Unknown Values

In Wilks's original representation, values which are unknown to an agent are represented by the agent not being a capable evaluator. In other words, awareness of the value of a λ-expression is represented by being able to evaluate the λ-expression. In figure 4 this would be equivalent to replacing the section below "*aware of the value of P*" by "*the correct value.*" Wilks's representation does not handle problems of awareness of P.

Being capable of evaluating the expression is still the criterion for knowing what the expression represents. Now, however, by the introduction of what we call *Uncertain Values* and *Uncertain_but_Incorrect Values*, we are able to express more with the representation.

An uncertain value is a value (of a CEV pair) which is unknown to an agent, i.e., the agent is aware that the value exists but not what the value is. This uncertain value may or may not be equal to a known value or some other uncertain value. Thus, the agent can represent the value that another agent believes is the value of a λ-expression without knowing what the value is.

Uncertain_but_incorrect values are known by the agent to be different from the value that the agent believes to be correct. In other words the value, while unknown to the agent, is believed by the agent to be wrong (the meta-beliefs that A believes that what B believes is the value of P is wrong, even though A does not know what B believes is the value of P).

An example is in order here. We annotate our representation by using ψ to represent an uncertain value and $\overline{\psi}$ to represent an uncertain_but_incorrect value:

$(PHONE\text{-}NO\text{-}OF\ Mary)\ Be\ (\lambda(x)\ (PHONE\text{-}NO\text{-}OF\ x)\ Mary)$

$$
\begin{bmatrix}
< \{Mary,\ System\}, v_0 >, \\
< \{Dan\}, v_1 >, \\
< \{Sally,\ Paul\}, \psi_0 >, \\
< \{Fred,\ John\}, \overline{\psi}_0 >
\end{bmatrix}
\tag{6}
$$

How do we interpret this structure? If this is a belief of the system then it may be interpreted as follows: The system believes that Mary and itself know the correct value[7] for Mary's phone number, namely v_0. Dan is believed to know Mary's phone number to be v_1, i.e., he knows the incorrect value, a value known to the system, but does not know that value to be incorrect. Sally and Paul both know the same phone number for Mary which may or may not be correct as the system does not know what they believe Mary's phone number to be. Finally Fred and John know Mary's phone number to be a value that the system does not know, however the system believes Fred and John to have the wrong phone number.

One problem of λ-formulas is that given some λ-expression in a nesting such that the agent at the bottom of the nesting can evaluate it, but an agent higher up in the nesting is unaware of the value of the expression, how then can the evaluation of the expression within the nesting be prevented?

For example, the λ-expression $(\lambda(x).P(x))\ [< \{Dan, System\},\ v_0 >]$, represents some function on x which Dan and the system are capable of evaluating, and resides in an environment for what the system believes Sally believes Dan believes. Sally is not a capable evaluator of the expression and so the expression should not be evaluable, even though it lies within an environment which has Dan as the innermost agent.

Should evaluation be prevented? It seems obvious that evaluation to a *specific value* should indeed be prevented. The representation allows for unknown values by means of uncertain values. What do the uncertain values represent? They are intended to allow the system to represent values that it doesn't know, but whose existence it is aware of. Thus the system is able to reason about these values without knowing what they are. In this case we wish to represent Sally being unaware of the value of P. A value called an *Unknown* value is used to represent this.

Unknown values are values which are unknown with respect to the environment in which they appear. Hence, if $(\lambda(x).P(x))\ [< \{Dan, System\},\ \mu >]$ lies within an environment which represents what the system believes Sally believes Dan believes, the expression evaluates to an unknown value (represented by μ).

We have demonstrated representations of all cases where agent B (figure 4) is believed to be aware of P. Next, we consider the case where B is unaware of P.

4.4. Awareness, and Explicit & Implicit Mention of Agents

A question is "*how to represent an agent being aware of the proposition represented by a λ-formula?*" Our solution is this: An agent is said to be aware of the proposition represented by a λ-formula if and only if that agent is either explicitly or implicitly mentioned in the set of all Δs for the λ-formula, otherwise the agent is said to be unaware of

---

[7]   Whichever set, if any, that the system lies within defines the "*correct*" value from the system's point of view.

the proposition.

In section 4.1 we say that a λ-formula consists of a λ-expression and a function table, and that the function table is a set of tuples $\Xi = (\ \Delta, \ \upsilon)$. The union of all such Δs for a given λ-formula, $(\overline{\Delta} = \bigcup_{all \ \Delta s \ in \ \lambda} \Delta \ )$, is the set of all capable evaluators for the given λ-formula. An agent is explicitly mentioned if the agent is a member of $\overline{\Delta}$ and is implicitly mentioned if the agent is a member of a class of agents[8] that is explicitly mentioned in $\overline{\Delta}$. This may be seen in example 5 where MDs are capable evaluators. Hence, any agent who is an MD is capable of evaluating example 5. MDs are explicitly mentioned in this example, someone who is an MD is implicitly mentioned.

An agent who is neither explicitly nor implicitly mentioned is deemed to be unaware of the proposition that the formula represents. In the case where an agent is implicitly mentioned in more than one class of agent, the most specific class is chosen to determine that agent's view.

It may appear that this representation requires the enumeration of countless number of agents. This is avoided by the use of classes of agents. If the majority of agents believe a particular λ-formula to return the value v_i, then this may be represented by having a Ξ with "average_man" as a member of its Δ, and with its υ equal to v_i. This is the least specific class of agents possible.

4.5. Function Transformations

The use of λ-formulas poses a question, what affect does ascribing a λ-formula to an agent have upon the λ-formula?

Ascribing a λ-formula often requires altering the formula. For example, ascribing a formula to an agent, who does not know the value it returns for any agent, will require changing the formula to reflect this situation.

Ascribing a λ-formula to an agent thus involves changing the function table for the formula. This is equivalent to saying that ascribing a λ-formula to an agent involves changing the function that the formula represents. Hence, we define a transformation function (Υ) which (for a given agent) maps from the domain of λ-formulas into λ-formulas.

$$\Upsilon: \ \delta \times \Lambda \rightarrow \Lambda'$$

or

$$\Upsilon: \ \delta \times \lambda \times \Gamma \rightarrow \lambda \times \Gamma'$$

where δ is an agent. In other words, the transformation function, Υ, maps a tuple (λ, Γ) into a tuple with the same λ but a different[9] Γ.

A function table Γ is a set of tuples $\Xi = (\ \Delta, \ \upsilon)$. The transformation function Υ will typically involve merging some of these Ξs and introducing new ones. The nature of the transformation function Υ is under investigation. We are currently investigating the role that the semantics of the λ-expression plays in determining Υ. That the semantics of the λ-expression *does* play a role in Υ may readily be seen by considering two examples. Consider two expressions, one of which pertains to Mary's phone number, and the other to a facet of Mary's health. The heuristic assumptions that the Υ function makes is different in these two cases by virtue of the nature of the information that they contain. We can assume that Mary knows her own phone number but not that she accurately knows the state of her health.

---

[8] Meaning Postulates are used to determine if an agent is a member of a particular class of agents.

[9] The resultant Γ may be the same as the original.

4.6. λ-expressions and Meta-Beliefs

We can see how λ-formulas solve the problems which were discussed in section 3. Section 3.1 describes beliefs which block the ascription of other beliefs. The role of this class of beliefs is totally subsumed by λ-formulas as this class describes lack of awareness of some proposition; lack of awareness is realised within λ-formulas by an agent being neither an explicit nor implicit agent of the formula's capable evaluators set ($\overline{\Delta}$). The remainder of section 3 describes atypical beliefs. Atypical beliefs are characterised by their associated meta-beliefs. λ-formulas are very effective at representing meta-beliefs of atypical beliefs.

While it may appear at first glance that the introduction of λ-formulas has created more problems than it has solved, this is not true. The problem of representing atypical and expert belief is a major problem compared to the problems presented by λ-formulas. Furthermore, the problems presented by λ-formulas have provided important insights into belief systems and the nature of meta-beliefs of atypical beliefs.

5. Conclusions[10]

A feasible belief system requires mechanisms to allow it to reason about the beliefs of other agents, without prestoring such beliefs. *ViewGen* is a system with just such mechanisms.

In this paper we have shown some of the problems involved in ascribing beliefs to agents, having limited knowledge of the agents' beliefs prestored. Further, we have demonstrated how the use of λ-formulas may circumvent many of these problems.

ViewGen a system written in Prolog is currently under development. It is intended that it be used as a module to perform certain forms of pragmatic reasoning, and by systems for natural language comprehension, dialogue analysis, planning, and human-machine interaction.

This work differs from that of (Konolige, 1984), (Levesque, 1984), (Haas, 1986), and (Maida, 1986) in that, while they provide languages to represent beliefs, we go further by providing heuristics for the ascription of belief. So, in our system, we actually have the function tables for λ-formulas and the Υ function to transform these formulas.

We are currently investigating the nature of the transformation function described in section 4.5 and the role that meta-beliefs play in reasoning with beliefs. We hope to be able to report on these topics in a future paper.

---

[10] Acknowledgements: My thanks to Yorick Wilks for his contribution to and supervision of this work. I would also like to thank Dan Fass for helping me straighten out the taxonomy of meta-belief.

6. Bibliography

Allen, J.F. & Perrault, C.R. (1978) Participating in Dialogues Understanding via Plan Deduction. *Proc. of the 2nd National Conference, Canadian Society for Computational Studies of Intelligence*, Toronto.

Ballim, A. (1986) Generating Nested Points of View. *Memoranda in Computer and Cognitive Science*, MCCS-86-68, Computing Research Laboratory, New Mexico State University, Las Cruces, NM 88003, USA.

Cohen, P.R. (1978) On Knowing What to Say: Planning Speech Acts. Technical Report No. 118, Dept. of Computer Science, University of Toronto.

Haas, A. (1986) A Syntactic Theory of Belief and Action. *Artificial Intelligence* 28, 245-292.

Konolige, K. (1984) Belief and Incompleteness. SRI Report No. 319.

Konolige, K. (1985) A Computational Theory of Belief Introspection. *Proceedings of IJCAI-85*, 502-508. International Joint Conference of Artificial Intelligence.

Levesque, H. J. (1984) A Logic of Implicit and Explicit Belief. *Proceedings of the National Conference on Artificial Intelligence*, 198-202.

Maida, A. S. (1986) Introspection and reasoning about the Beliefs of other Agents. *Proceedings of Cognitive Science Society*, 187-195.

Moore, R.C. (1980) Reasoning About Knowledge and Action. Artificial Intelligence Center Technical Note 191, SRI International.

Weyhrauch, R. (1980) Prologomena to a Theory of Formal Reasoning. *Artificial Intelligence* 13, 133-170.

Wilks, Y. (1986). *CRL Work on Beliefs and Computation.* Paper presented at the Workshop on the Foundations of Artificial Intelligence, University of Naples, Naples, Italy.

Wilks, Y. & Ballim, A. (in press) The Heuristic Ascription of Belief. In *Progress in Cognitive Science, Vol II.*, N. Sharkey (ed.), Erlbaum.

Wilks, Y. & Bien, J. (1979) Speech Acts and Multiple Environments. *Proceedings of IJCAI-79*, 451-455. International Joint Conference on Artificial Intelligence.

Wilks, Y. & Bien, J. (1983) Beliefs, Points of View and Multiple Environments. *Cognitive Science* 8:120-146.

KNOWING THAT AND KNOWING WHAT

Allan Ramsay

Cognitive Studies Programme
University of Sussex, Falmer BN1 9QN

ABSTRACT

This paper presents a set of inference rules for reasoning about other people's knowledge. These rules are not complete, but they are effective. The paper shows how they may be used with a non-trivial set of contingent facts and rules about the knowledge possessed by various people to answer fairly complex queries about who knows what. The inference rules are based on the notion of mimicking the reasoning that might be performed by another person, rather than on an analysis of the conditions under which some fact is constrained by the other facts which may be available. The conclusion argues that the speed of the resulting system at least partially compensates for the fact that it does not manage to derive every possible conclusion.

INTRODUCTION

Consider the following situation: Allan knows that Alex and Trish live together. He also knows that Janet knows Alex's phone number. How does he know that asking Janet what Alex's number is will enable him to phone Trish?

This problem is typical of the sort of reasoning we have to perform every day in order to function in the world. We need to be able to reason not just with our own knowledge about the world, but with other people's. We need it in order to be able to work co-operatively with other people, since we have to be able to predict what information they will need in order to carry out their part of the task. We need it to be able to work against them, so that we can set them traps (if, for instance, you were playing someone at chess you would not try to catch them with a fool's mate if you knew they already knew about it). We need it in order to be able to carry out satisfactory dialogues with other people - if you assume they know more than they in fact do then they will fail to understand you, if you assume that they know less then they will get bored.

Reasoning about other people's knowledge is recognised in AI as a major task. A number of people have proposed ways of dealing with it, notably Moore (1984) and Konolige (1982). These solutions tend to be elegant but computationally intractable. Moore (1980), for instance, explicitly acknowledges that at that time he had no working theorem prover for his treatment of the problem, and the later paper gives little evidence of one either. Appelt (1985) does use a theorem prover based on Moore's analysis of the problem, in order to reason

279

about what it may be appropriate to say when conversing with a novice in some domain. The performance of this system indicates that although an implementation exists, there is considerable room for improvement in terms of time taken to reach conclusions.

This paper presents an alternative scheme for reasoning about other people's knowledge. The mechanism described below is incomplete, in that there may well be things which follow from what you know about another person's knowledge but which this system cannot infer. It is, however, fast enough to be included within, for instance, a natural language generation system without completely swamping the processing done by the rest of the system. The present implementation contains rules for analysing problems like the one that introduced this section of the paper.

OBSERVATIONS ON REASONING ABOUT KNOWLEDGE

The following points are critical for any attempt to reason about other people's knowledge (these points are well-known, and I claim no credit for pointing them out here. Nonetheless they need to be restated).

(i) No-one can know anything that is not true. As a direct consequence, if A knows that B knows P, then A must know P. Knowledge thus differs from belief, since there is no reason to suppose that if A believes that B believes P then A will also believe it. The distinction is slightly artificial, since no-one ever really *knows* anything, except perhaps logical tautologies. All we ever have is very strong beliefs. Nonetheless, people and AI planning systems frequently behave as though they thought they did know things. The task of recovering when what was thought to be known turns out to be wrong is, of course, another open problem in AI. We will follow all other AI workers on knowledge about knowledge by assuming that someone else will solve the truth maintenance problem for us. The current system assumes that statements about its own and other people's knowledge are indeed correct, and are not just beliefs.

(ii) In general, then, knowing that someone else knows something is grounds for inferring it for yourself. There is one very specific case in which it fails. If I overhear Janet saying "It's raining", then I can infer that it is indeed raining (subject to all the usual restrictions about the correctness of Janet's belief). From this I can, if I want, infer that I know it's raining, or do whatever else I want with it. If I overhear her saying "I know Alex's phone number" then all I can infer is that Alex has a phone number. In case (i), I know exactly what Janet knows as soon as I know she knows it. In case (ii) I know less than she does.

(iii) The mechanism described in (i) is transitive - if A knows that B knows that C knows that P then A can infer P. This chaining of knowledge about people's knowledge can be carried out as many times as you like.

(iv) It is generally assumed that everyone is capable of performing the same sets of inferences. This is yet another idealisation, though it is not usually pointed out with as much care as the others that are made in this area. If we can assume this, then much of the difficulty of dealing with knowledge about knowledge disappears. In order to see whether someone else is capable of drawing some conclusion all we need to do is see how we would draw it ourselves, and then check that we have not used any contingent facts or rules which we do not know are available to them.

This approach can be extended to deal with people whose ability to reason is inferior to our own, simply by regarding some or all of our rules of inference as contingent rules to which they do not have access. It cannot be extended to people whose reasoning is superior to our own, but then no approach can possibly succeed in that. It is not possible for A to know that B has access to some rule, say R1:(P1 & P2 ... &Pn) → Q, without having access to it himself. A may know that B has access to some rule R1 whose form he doesn't know; but in that case he will not be able to reason about the conclusions B may draw by using it. (This one does not go away if A knows a bit about the form of B's inference rule, but not everything. Suppose, for example, A knows that B's rule is either R1:(P1 & ... &Pn) → Q1 or R2: (P1 & ... & Pn) → Q2. Then if A knows that B knows P1 and P2 and ... Pn, then he knows that B knows Q1 or Q2, and hence can infer for himself that Q1 OR Q2 is true. But this is different from, and less than, what B knows - A just knows that Q1 OR Q2 is true, whereas B really does either know that Q1 is true or know that Q2 is). There is a final case we might need to consider when worrying about other people's ability to reason, namely when their reasoning is not better or worse than our own, but different in some other way. It might be that their reasoning differs from ours by employing a different, but equally valid, set of axioms and rules of inference. If we knew this, and knew what they were, we could again take their axioms and rules to be contingent facts which we knew they had access to, and continue to mimic them. Since we are concerned with knowledge and not belief, there is no other way for their reasoning to differ from our own. If we know that they use a sound inference rule, and if furthermore we know what it is, then we can use it ourselves.

A consequence of the above discussion is that we can examine other people's knowledge simply by seeing what we ourselves can infer, and then checking we know that they know all the basic facts we have used. The remainder of the paper presents a set of inference rules, and a collection of contingent facts and rules about various people's knowledge, which will support the kind of reasoning needed for the introductory example. The inference rules are not complete. In particular, they do not support reasoning from hypotheses. It should be remembered that any set of inference rules must either be incomplete or run the risk of non-termination. The rules given below could easily be extended to cover extra cases as required.

RULES FOR REASONING ABOUT KNOWLEDGE

The following rules are written in a PROLOG-like notation, with variables indicated by a preceding @ sign, and the heads and bodies of clauses separated by "if". It is important to note that the inference engine used stops mutually recursive clauses from producing infinite loops (so that it is quite safe, for instance, to use rules like "X lives with Y if Y lives with X"). Full details and program text for this inference engine are given in Ramsay and Barrett (1987).

The rules are split into two sections. The first group are general rules of inference. The second group are contingent facts and rules, which are marked as being known by one or more people. The system uses the first group on the second to draw conclusions about various people's knowledge. It must be clearly understood that the reasoning system we are interested in consists of the inference rules plus the contingent facts, and that no contingent fact can be provided without the system knowing it - it just would not make sense.

Inference rules The following rules are used by the system for its own inferences. It is assumed, however, that the same set of rules are available to everyone else, so that anything that can be concluded using these rules and the facts available to some specific other person can be inferred by that person.

The rules make use of three conventions, as follow:

(i) an expression such as **knows(a, f, [b,c,d])** should be read as "D knows that C knows that B knows that A knows that F". In other words, the third argument to **knows** is used for nesting statements about what people know about each other's knowledge. This way of expressing such statements enables us to delay worrying about whether people really do know what we have hypothesised them to know until we need to check the particular facts they have access to. The effect could have been achieved by some rule which converted a nested expression to the form we want, for instance

knows(@A, knows(@B, @P)) if knows1(@B, @P, [@A]).

We leave the rule in the form which we actually want it in for simplicity.

(ii) Contingent facts and rules are represented by statements of the form

fact(@N, @X, @P)

where **N** is an index for the fact (so that it may be referred to elsewhere), **X** indicates who knows it, and **P** is the fact itself. For rules, **P** is a compound expression containing an implication arrow. Thus

fact(1, Janet, that(raining))

says that the first fact the system knows about is that Janet knows it is raining (we sometimes say below "'it is raining' is a fact for Janet"). The indices are not used for anything except cross-references between facts, for instance

fact(2, Allan, that(1))

could be used to indicate that the system knows that Allan knows that Janet knows it is raining. There must, clearly, be a way to represent facts which are available only to the system. The simplest way to do this is to permit an identifier such as "I" to stand for the system. It does not seem as though there is any need for specific rules about "I". The example below does not refer to "I". It is also necessary to allow facts to be known to more than one person. The simple indexing scheme used here makes it possible to refer to facts and rules that are known by everyone. For instance,

fact(3, @ALL, that(lives_with(@X, @Y)) → that(lives_with(@Y, @X)))

says that everyone knows that if X lives with Y then Y lives with X. As the system stands, facts which are known to a specific finite group of people can be recorded by enumerating them, as in

fact(3a, Allan, that(raining))
fact(3b, Julian, that(raining))
fact(3c, Ros, that(raining))
...

Rules which are true for some generic group of people, such as "all logicians know that automatic theorem proving for modal logic is hard", cannot easily be stated with the formalism as it stands. This is an area which needs more work.

(iii) Contingent facts are divided into 'identifying' and 'non-identifying' facts. Identifying facts are facts which state that someone not only knows that something is true, they also know how to refer to the things it is true of. Such facts are indicated by the use of the quantifier **what**, as in:

fact(4, Janet, what(sk1, phone_number(Alex, sk1)))

The notation is not perhaps as neat as it might be, but the general idea should be clear. The identifiers **sk1, sk2,** ... are used as Skolem constants to indicate existential quantification, the quantifier what is used to show that the person concerned has a rigid designator for the the thing referred to. Non-identifying facts are marked by the prefix **that**, so that

fact(5, Allan, that(lives_at(Alex, sk2)))

would indicate that Allan knew Alex lived somewhere, but that he did not know where.

Finally we come to the inference rules themselves. They are each prefaced by an English gloss. This gloss frequently starts with a parenthesis referring to a list of people. This is the list referred to in note (i) above with respect to whom the given rule is taken to operate.

Rule 1: (The people listed in ALL_KNOW know that) X knows P if P is a fact for X and they all know it is:

knows(@X, @P, @ALL_KNOW)
if
fact(@N, @X, @P),
check(@N, @ALL_KNOW).

This is used with facts like 1 and 2 above to deal with questions like "Does Allan know that Janet knows it's raining?". **check** is a predicate, defined below, which finds out whether people have direct access to particular facts.

Rule 2: (The people listed in ALL_KNOW know that) X knows P and Q if they know that X knows P and they know that X knows Q:

knows(@X, and(@P, @Q), @ALL_KNOW)
if
knows(@X, @P, @ALL_KNOW),
knows(@X, @Q, @ALL_KNOW).

This one simply reflects the assumption that everyone knows that if two things are both true then their conjunction is true. Since everyone knows it, it can be used at any point when we want to mimic someone else's reasoning.

Rule 3:   (The people listed in ALL_KNOW know that) X knows Q if they
know that X has access to a rule of the form "P implies Q" and they also
know that X knows P:

> knows(@X, @Q, @ALL_KNOW)
> if
> fact(@N, @X, @P → @Q),
> check(@N, @ALL_KNOW),
> knows(@X, @P, @ALL_KNOW).

This reflects the assumption that everyone knows modus ponens, i.e. that if you
know P → Q and you know P then you can infer Q. It is important to note
that although the rule allows the system to try to infer the antecedent using
whatever means are available to it, the implication itself must be a fIfactfR.
This is one of the places where we have chosen not to give the system the
strongest possible inference rule - the given rule does support a lot of the ways
that people use implications (for instance, that from P, P → Q and Q → R you
can infer R), but it does not support arbitrary attempts to infer new contingent
rules. This is a deliberate choice which could easily be altered. Allowing the
system to try to draw its own inferences would have enabled it to infer a few
more consequences from whatever set of contingent facts and rules it was given,
at the cost of a massive degradation of performance.

Rule 4:   (The people listed in ALL_KNOW know that) X knows that X knows
F if they know that X knows F. (This is an idealisation. The only realistic
way to restrict its application is to put a resource limitation on the inference
process. This could easily be done, but it would add little to the interest of the
example, and would just make things harder to follow):

> knows(@X, knows(@X, @F, @ALL_KNOW), @ALL_KNOW)
> if
> knows(@X, @F, @ALL_KNOW).

This rule is allowed in by most people working in this area, to reflect the fact
that you ought to be able to introspect on what you know. It is not used in
any of the examples given below, but it is available for those few problems
where it is actually useful.

Rule 5 (The people listed in ALL_KNOW know that) X knows that Y knows
F if the list of people we get by adding X to ALL_KNOW know that Y
knows F:

> knows(@X, knows(@Y, @F, @ALL_KNOW), @ALL_KNOW)
> if
> knows(@Y, @F, [@X | @ALL_KNOW]).

This rule allows us to delay worrying about how Y might infer F. All the
other inference rules simply pass ALL_KNOW on unchanged until a specific
contingent fact or rule is required. We thus do not need to check that everyone
knows that Y can perform inferences. All that has to be checked is that he is
known to have access to the specific contingencies.

Rule 6: (The people listed in ALL_KNOW know that) X knows P, where P is a non-identifying proposition, if there is some other person, Y, who knows P, and the people in ALL_KNOW know that X knows Y knows it:

> knows(@X, that(@P), @ALL_KNOW)
> if
> fact(@N, @Y, that(@P)),
> check(@N, [@X | @ALL_KNOW]).

This comes from the fact discussed above that no-one can know anything which is untrue, so that if X knows that Y knows P then X must be able to infer P.

Rule 7: (The people listed in ALL_KNOW know that) X knows P, where P is a non-identifying proposition, if there is some other person, Y, who knows an identifying proposition about P, and the people in ALL_KNOW know that X knows that Y knows it:

> knows(@X, that(@P), @ALL_KNOW)
> if
> fact(@N, @Y, what(@W, @P)),
> check(@N, [@X | @ALL_KNOW]).

This is aimed at the case where Allan knows that Janet knows what Alex's phone number is. We cannot afford to derive the conclusion that Allan also knows what it is, but we do want to be able to infer that he at least knows Alex has a phone.

The final pair of rules discharge the nested collections of people who know about each other's knowledge. The first rule says that if there are no people in the list then there is nothing to be done. The second simply unwinds the nest so that explicit statements of relevant facts are searched for:

> check(@N, []).

> check(@N, [@C | @ALL_KNOW])
> if
> knows(@C, that(@N), @ALL_KNOW).

<u>Contingent facts and rules - an example</u> The first four facts refer to the specific knowledge possessed by Allan and Janet. They say that Allan knows that Alex and Trish live together; that Janet knows Alex's phone number; that Allan knows Janet knows Alex's phone number (but not that he knows it himself); and that Allan knows Janet's phone number:

> fact(1, Allan, that(lives_with(Alex, Trish))).
> fact(2, Janet, what(sk1, phone_number(Alex, sk1))).
> fact(3, Allan, that(2)).
> fact(4, Allan, what(sk2, phone_number(Janet, sk2))).

The next group of facts express various pieces of common knowledge. They can be used in chains of inference about the knowledge possessed by anyone at all, since the place for the name of the individual possessing the knowledge is filled by a variable (which will match the name of any specified individual):

Fact 5: (Everyone knows that) anyone who has a phone knows its number:

```
fact(5,@ALL,
      that(phone_number(@P,@N))
          →knows(@P,what(@N,phone_number(@P,@N)))).
```

Fact 6: (Everyone knows that) if two people live together and one of them has a phone, the other one has a phone with the same number:

```
fact(6, @ALL,
      (that(lives_with(@X, @Y)) and that(phone_number(@X, @N)))
          → that(phone_number(@Y, @N)))).
```

Fact 7: (Everyone knows that) if Y knows P then if X asks Y about P then X will know P as well:

```
fact(7, @ALL,
      knows(@Y, @P, @ALL_KNOW)
          → effect(ask(@X, @Y, @P), knows(@X, @P, @ALL_KNOW)))).
```

It is assumed here that Y is perfectly co-operative.

Fact 8: (Everyone knows that) if X lives with Y then Y lives with X:

```
fact(8, @ALL,
      that(lives_with(@X, @Y)) → that(lives_with(@Y, @X)))).
```

Note that this rule will lead to infinite regress unless the inference engine contains an explicit check.

Fact 9: (Everyone knows that) if X lives with Y then X knows X lives with Y:

```
fact(9, @ALL,
      that(lives_with(@X, @Y))
          → knows(@X, that(lives_with(@X, @Y)), [])).
```

Fact 10: (Everyone knows that) if X can call Y then X can ask Y about anything he likes:

```
fact(10, @ALL,
      can(@X, call(@X, @Y)) → can(@X, ask(@X, @Y, @ABOUT)))).
```

Fact 11: (Everyone knows that) if X knows Y 's phone number he can call Y:

```
fact(11, @ALL,
      knows(@X, what(@N, phone_number(@Y, @N)), []
          → can(@X, call(@X, @Y))))).
```

This rule set contains a number of bits of common knowledge, among which are a number of rules which give the preconditions and effects of the actions "calling" and "asking". These rules are clearly a long way from complete characterisations of these rather complex actions. It is also evident that while descriptions of actions in terms of effects and preconditions are useful to the standard AI planning algorithms, we have not given any implementation of a planner. In particular, the reasoning mechanisms outlined here are inadequate for investigating the effects sequences of more than one action. Moore (1984) does manage to mix chains of reasoning about other people's knowledge with investigations of the effects of sequences of actions, but, as noted above, his system does not seem to provide the performance that ours does. We would prefer to solve the problem of complex sequences of actions by adding to our system some mechanism for simulating the effects of the actions. Nonetheless, the ability of the system to answer the questions given below indicates that something like the current system might well by very useful for reasoning about how to plan to ask somebody something.

Examples of questions the system can answer   As with the rule set, the examples are prefaced by English glosses. We do not pretend to have any programs which can translate in either direction between English and the notation used by the reasoning system. Readers who do not believe that there is any close correspondence between the formal notation and what we are offering as an English equivalent should disregard the glosses. We regret that we will probably not be able to convince such a reader that we have anything interesting to offer.

Question: whose phone number does Allan know?
Answer: Janet's.

GOAL(knows(Allan, what(@N, phone_number(@X, @N)), [])).

ANSWER:
{knows Allan {what sk2 {phone_number Janet sk2}}} [].

Note that both the query and the answer have [] as their final component, indicating that the query was directly about what Allan knows, rather than about who else knows something about Allan's knowledge.

Question: what action does Allan know would lead to his knowing Alex's phone number?
Answer: he would know it if he asked Janet what it was.

GOAL(knows(Allan,
        effect(@ACTION(@X, @Y, @Z),
        knows(Allan, what(@N, (phone_number(Alex, @N)), [])))), [])

ANSWER:
{knows/4 Allan
        {effect {ask Allan Janet {what sk1 {phone_number Alex sk1}}}
               {knows Allan {what sk1 {phone_number Alex sk1}} []}}
        []}

Question: what action does Allan know would lead to his knowing Trish's phone number?
Answer: he could ask Janet what Alex's phone number was. The chain of inference that leads to this conclusion depends on knowing that since Trish and Alex live together they must have the same phone number, so anyone who knows Alex's phone number will be able to provide the required information. Note that there is no need for Janet to know that Trish and Alex live together

for this inference to work:

GOAL(knows(Allan,
        and (that(phone_number(Trish, @N)),
            effect(@ACTION(@X, @Y, @Z),
                knows(Allan, what(@N, @E), [])), []))
ANSWER:
{knows/4 Allan
        {and {that {phone_number Trish sk1}}
            {effect {ask Allan Janet {what sk1 {phone_number Alex sk1}}}
                {knows Allan
                            {what sk1 {phone_number Alex sk1}}
                            []}}} []}

This example should be taken as indication of the point at which the system
starts to break down. The English gloss said "what action does Allan know
would lead to his knowing Trish's phone number?". The query that was
actually asked was more like "Does Allan know that Trish has a phone number
and that there is some action which would enable him to find it out?" This is a
consequence of our choice of inference rules. Stronger rules would have enabled
the system to answer the right question. Note however that what Allan is to
ask Janet is {what sk1 {phone_number Alex sk1}}, which can easily be glossed
as "What's Alex's phone number?". This question correctly avoids mentioning
Trish's number, since the available facts give no indication that Janet has ever
heard of Trish, let alone that she knows she has the same phone number as
Alex.

Question: what action does Allan know that he can perform which will lead to
his knowing Trish's phone number?
Answer: he knows that he can phone Janet and ask her Alex's number. This is
more complex than the previous case, since Allan has to prove not only that
Janet will provide the information if asked, but also that he can ask her
because he knows her number and hence can ring her:

GOAL(knows(Allan,
        and (and (that(phone_number(Trish, @N)),
            effect(@ACTION(@X, @Y, @Z),
                knows(Allan,
                    what(@N, phone_number(Trish, @N), [])))),
            can(Allan, @ACTION(@X, @Y, @Z))), []))

ANSWER:
{knows/4 Allan
        {and {and {that {phone_number Trish sk1}}
                {effect {ask Allan Janet
                                {what sk1 {phone_number Alex sk1}}}
                            {knows Allan
                                {what sk1 {phone_number Alex sk1}}
                                []}}}
            {can Allan
                {ask Allan Janet {what sk1 {phone_number Alex sk1}}}}}}
            []}

CONCLUSIONS

Previous workers in this area have concerned themselves largely with systems which will be able to draw all possible inferences about the states of knowledge of the various participants. Such systems have rather more power than the rule set given above, as is indicated by our last two examples, where the questions have had to be phrased extremely carefully. Nonetheless, our system has one great advantage: it is quick. Despite the fact that the inference engine which is used for finding and applying rules is, by the standards of most PROLOG compilers, rather slow (around 200LIPS on a VAX_11/750, compared to 5KLIPS upwards for commercially available compilers), the answers to the example queries took less than 10 seconds. Extending the set of inference rules would be expected to worsen the performance, but with finer tuning the background inference engine could be brought up nearer the 2KLIPS mark (it will always be slower than a pure PROLOG engine, since it needs to check for infinite loops). We cannot, in any case, ever afford to use a complete theorem prover in any autonomous or semi-autonomous system, since there will always be a risk of non-termination. It seems, therefore, entirely reasonable to aim for a system which will answer quite a lot of the questions we might want to ask, and doesn't take too long about it.

The main problems with theorem proving with modal logic arise from the presence of what we have referred to here as 'identifying facts'. If it were not for situations such as Allan knowing that Janet knows what Alex's phone number is without knowing it himself, the task would scarcely be more problematic than theorem proving for ordinary logic. The introduction of the quantifier **what** makes some of the problems more tractable. We have not, however, given much analysis of the *meaning* of this quantifier. It is easy enough to understand what it means to say "I know what Alex's phone number is" - it means that I know a sequence of numbers which when dialled will enable me to ring him up. The conditions under which I can truly say "I know who Alex's girlfriend is" are far less clear - does knowing her name suffice, or knowing what she looks like, or knowing that she is a lecturer at St Martin's School of Art? It is quite conceivable that I might know "who someone is" without ever having met them and without knowing their name, or what they look like, or indeed any true fact about them. The philosophical literature does not seem to have any very solid analysis of this problem. Most of the discussion has centred on the problem of "knowing who or what a name refers to", but an answer to this problem would probably provide an answer to our difficulty about "knowing who or what a definite reference refers to". Kripke's (1971) introduction of 'rigid designators' accounts for cases like phone numbers. As far as other entities are concerned, we leave it as an open problem. At any rate, even without a proper analysis of when we do and do not know what something is, we can be sure that our inference rule 7 (if X knows that Y knows **what(@W, @P)** then X knows **that(@P)**) is sound.

References

Appelt, D., 1985. Planning English sentences Cambridge University Press, Cambridge.
Konolige, K., 1982. A first-order formalisation of knowledge and action for a multi-agent planning system, in Machine Intelligence 10, (eds. Hayes, Michie and Pao), Ellis Horwood, Chichester.

Kripke, S.A., 1971. Semantical considerations on modal logic, in Reference and modality (ed Linsky), OUP, London.

Moore, R.C., 1980. Reasoning about knowledge and action SRI AI Center Report 191.

Moore, R.C., 1984. A formal theory of knowledge and action, in Formal theories of the commonsense world (eds. Hobbs and Moore), Ablex Pub.Corp., New Jersey.

Ramsay, A.M. and Barret, M.R., 1987. AI in practice: examples in POP-11 Ellis Horwood, Chichester.